SpringerBriefs in Philosophy

More information about this series at http://www.springer.com/series/10082

Chelsea C. Harry

Chronos in Aristotle's Physics

On the Nature of Time

Springer

Chelsea C. Harry
Philosophy Department
Southern Connecticut State University
New Haven, CT
USA

ISSN 2211-4548   ISSN 2211-4556 (electronic)
SpringerBriefs in Philosophy
ISBN 978-3-319-17833-2   ISBN 978-3-319-17834-9 (eBook)
DOI 10.1007/978-3-319-17834-9

Library of Congress Control Number: 2015936467

Springer Cham Heidelberg New York Dordrecht London

Printed on acid-free paper

Springer International Publishing AG Switzerland is part of Springer Science+Business Media
(www.springer.com)

For TS

Preface

> *Every realm of nature is marvelous: and as Heraclitus, when the strangers who came to visit him found him warming himself at the furnace in the kitchen and hesitated to go in, is reported to have bidden them not to be afraid to enter, as even in that kitchen divinities were present, so we should venture on the study of every kind of animal without distaste; for each and all will reveal to us something natural and something beautiful*
>
> (645a17–24).

Philosophy, as it is practiced in contemporary Western academia, is overwhelmingly problem based. Thought experiments take our search for truth outside of the context in which it came to be a search at all. In light of this commitment, we deemphasize context—it is always only accidental.

To assuage this commitment, or, at the least, call it into question, frees us to talk about the place of context and history in our pursuit of wisdom. When we combine a search for truth about a problem with a historical approach, we use a text as a referent for what *was real* for the author. To do justice to the author's understanding is thus to bear witness to a double sense of context—it is an uncovering of a way of past thinking—whether or not the same type of thinking still holds today—in addition to a sorting through of present arguments shrouded in a certain organization; the arguments are always already *with a text*.

Reading the ancients takes us to another time, to another place: to a particular way of thinking about the world, to a specific way of experiencing life. When we turn to the ancients for wisdom, therefore, we must guard ourselves from extrapolating that wisdom from that which gives it meaning. With this said, I will discuss briefly my initial interest in understanding Aristotle's position on time (*chrónos*).

My first introduction to Aristotle's Treatise on Time (*Physics* iv 10–14) was to the arguments, the puzzles, or *aporiai* and their subsequent examination, excerpted from their context. The editors of a textbook about the issue of time presented Aristotle on time as four pages, given in translation and without introduction or annotation. My intuition then, confirmed years later, was that when taken from the wider *Physics*, divorced from Aristotle's natural philosophy, the treatise does not make sense. As many have before me, I found the treatise inconsistent and littered with jargon. Without imposing on Aristotle everything my modern mind understands about time and space, it was difficult to begin a serious study of his arguments.

Instead of adopting Aristotle's questions about time as my own, I decided to pursue a line of questioning that would help me to understand *why* these were his questions—both in the sense that I wanted to know his general method and also in the sense that I wanted to know why he was interested in time. This required not only a reading of the rest of *Physics* iv, but a study of the rest of the *Physics*, and indeed, much of Aristotle's works in natural philosophy and logic.

It soon became apparent to me that Aristotle was not interested in time in the sense that Newton or Einstein were interested in time. Aristotle was interested in time because he was interested in change, and change only because change for Aristotle is the nature of life. This is to say that Aristotle was interested in time only because he was interested in nature, in natural beings.

Aristotle's interest in nature led me to more questions still: What did he mean by 'nature'? What was his experience of the natural world? How did he understand humans in the context of life generally? Responding to these questions is an ongoing pursuit for me, an expedition that has brought me out of Aristotle's extant work and to the places where he lived, experienced, thought, and wrote. In particular, time I spent on the shores of the Bay of Kalloni, on the Island of Lesvos in the northeast Aegean, the place where Aristotle and Theophrastus inaugurated biological study, afforded me the opportunity to perceive life in the manner I imagine Aristotle must have perceived it—in its majesty, as that which was both the same as me and also other than me. Aesthetics, from the Greek *aesthesis*, came to mean beauty, or art, in the eighteenth century. The Greek *aisthesis*, αἴσθησις, can be rendered both as perception and feeling. Aistheton, αἰσθητόx, the perceptible object, was received, integrated, perhaps considered, felt. It was taken in.

Aristotle took in nature; he took in life. We celebrate him because he was the first to systematize this taking in; he sought not just to experience nature, but to know it—to categorize the "same" and "the different," to name its purpose. As any biologist or naturalist will tell you, the type of dedication it took for him not only to conduct the exacting and detailed studies of natural objects, but to do so when biological study was considered useless, even disgusting (see the invitation to biology in *PA* i 5), points to the conclusion that Aristotle had taken in fully the natural world. His consequent appreciation for its being and diversity resulted in the most prolific body of scientific writing penned by any one person.

It is with this in mind that I undertook the project to understand Aristotle's Treatise on Time. Aristotle came to time because he came to nature; he came to nature because he had a certain orientation to life, to his context. It is out of reverence for this orientation, as an extension of a genuine scholarly desire to know the truth, that I offer my reader the following interpretation of Aristotle's theory of time.

Thessaloniki, Greece   Chelsea C. Harry
New Haven, CT, USA

Acknowledgments

I will never be able to communicate appropriately or fully my immense gratitude to the people, places, and entities who helped me to see this manuscript to its current state of completion.

This manuscript is a revision of a portion of my doctoral dissertation, *Time (Chronos) in Aristotle's Natural Philosophy and of Time's Place in Early Naturphilosophie (1750–1800)*, written under the direction of Ron Polanksy at Duquesne University in Pittsburgh, PA. Without the guidance, support, and encouragement of Ron Polansky, Lanei Rodemeyer, and Gottfried Heinemann, the dissertation, let alone this revision, would not have been possible.

In the summer of 2013, Gottfried Heinemann spent a generous amount of time discussing with me my reading of Aristotle's *Physics*. He and his wife, Marion, welcomed me into their home in Kassel, Germany, showing me hospitality I will not soon forget.

My dear friend Justin Habash read the original dissertation despite having absolutely no obligation to do so. The boost I received from this single gesture of kindness is incalculable. And, his comments about the work had an undoubtedly positive effect on its revision.

I benefitted greatly from conversations had and feedback received from participants at the following conferences, where drafts of chapters at their various stages of completion were presented: The Society for Ancient Greek Philosophy (SAGP) in NYC (2011–2014); ATINER conference in Athens, Greece (2011); 5th Annual International Symposium on Aristotle: On the Issue of Time in Thessaloniki, Greece (2012).

A small faculty research group, which met during the summer of 2013 was the first opportunity I had to suss out some of the changes I wanted to make to the second chapter of the present work. I thank Rex Gilliland and Rafael Hernandez for their participation and friendship.

A CSU-AAUP research grant and an invitation from Prof. Demetra Sfendoni-Mentzou, President of the Interdisciplinary Centre for Aristotle Studies at the Aristotle University of Thessaloniki in Thessaloniki, Greece, permitted me a six-week research stay in Greece during the summer of 2014 as a Visiting Fellow at the

Centre. Ευχαριστώ Πάρα πολύ to the "team": Dimitra Balla, Christina Papachristou, Maria Kechagia, Tasos Zisis, George Stremplis, and Christina Gkountona. Thanks also to Professor Sfendoni's graduate students, especially Lia Rotsia, for attending my guest lecture on Aristotle's concept of time.

   Jason Tipton, the once anonymous reviewer for Springer, has been a source of encouragement and offered me many excellent suggestions to improve the third chapter of the present work. Likewise, I owe thanks to Lucy Fleet, acquisition editor for Springer, for her support of this project from my first contact with her in May 2013.

   Research reassigned time granted by SCSU and the Dean of Arts and Sciences, Steven Breese, during the 2014–2015 academic year, permission from Department Chair, Armen Marsoobian, to apply for research reassigned time, and administrative help from Department Secretary, Sheila Magnotti, allowed me requisite time and resources to complete revisions to this manuscript. The students in my Ancient Philosophy courses at SCSU have been an ongoing source of inspiration for me.

   I am indebted to the following individuals for contributions not only to this work, but to my life, big and small: Kelsey Ward, Becky Vartabedian, Bob Vallier, Tamara Albertini, David Pettigrew, Jim Swindal, Joan Thompson, Becky Day, Dana Ferrer, The Toledos, Diane Miller, Kyla Dennigan, Nolen Gertz, Tamara Albertini, Christina Takouda, Mieke de Moor, Ken Alba, Jackson, and Wolfgang Walsh.

   My debts to my family are the greatest, especially to my parents, Karen and Tim Harry, and to my husband, Chris Walsh.

   Though the arguments in this manuscript have been made better from so many helpful criticisms and from the insights of minds greater than my own, I alone take responsibility for the ways they inevitably fail.

Contents

1   **Time in Context**   1
     1.1   *Physics*: Scope, Access, Goals, and Method   2
     1.2   Nature and Its *Archai*   6
     1.3   The Role of *Kinêsis* in Nature   16
     1.4   From *Kinêsis* to *Chrónos*   22

2   *Physics* **Iv 10-11 as a Parallel Account**   33
     2.1   Introducing the Issue of Time   35
     2.2   Eschewing the *Endoxa*   37
     2.3   Aristotle's Positive Account of Time   40

3   **Taking Time**   51
     3.1   Conditions for Actualized Time   52
     3.2   Readiness for Thinking: From Marking to Counting   56
     3.3   Perceiving Time Revisitied   62

Concluding Remarks   69

Bibliography   71

Index   75

Abstract

Chronos *in Aristotle's Physics: On the Nature of Time* argues that Arisotle's Treatise on Time (*Physics* iv 10–14) is a highly contextualized account of time in so far as it is not a treatment of time *qua* time but a parallel account to Aristotle's foregoing studies of nature, principles (192b13–22), motion (201a10–11), infinite (iii 4–8), place (iv 1–5), and void (iv 6–9) in the *Physics* i–iv 9. It offers a reading of *Physics* iv 10–11 with the aim of showing that time, *chrónos*, here has to do with time as an attribute of motion, as an interval, i.e., the type of time that, as Aristotle describes at 218a1, "is taken." With support from a reading of *Physics* iv 14 and evidence from Aristotle's greater philosophy of nature, it argues that time for Aristotle is derivative of the modal change of natural being. Time is then only ever potentially actual unless this change is apprehended, in most cases, by the working together of perception and intellection and, in some cases, by perception alone. Studies in contemporary animal science help to buttress this final conclusion.

Keywords Aristotle · Nature · Time · Potentiality · Context · Natural beings · Scope · Method · Now · Continuity · Taking time · Analytic of time · Non-human animals · Soul · Perception · Subitizing · Counting · Number · Nous · Psyche

Introduction

The Study of Nature and the Nature of Time

> Aristotle's Physics is the hidden, and therefore never adequately studied, foundational book of Western philosophy.[1]

There was a time in history when it was difficult to read Aristotle's arguments in the *Physics* contextually; in the medieval and early renaissance periods, this was due to a lack of any standard or widely read translation of key works like the *Physics*. Even Aquinas's great commentary on the work was most likely completed using a *kontamination*, or mixture of various translations (Aquinas 1961, xix). In the late Renaissance, there was perhaps resistance to reading Aristotle's natural philosophy objectively; as Galileo argued in the sixteenth century, readers of Aristotle have been more keen to defend his reputation than they have been to adjudicate fairly the conclusions he drew.[2] Despite that Aristotle's *Physics* was nevertheless the dominant work on natural science in the West until the dawn of modern science, once the paradigm of nature for which it argued was supplanted in the seventeenth century—ultimately by the Newtonian model,[3] but Newton's work was helped

---

[1]Martin Heidegger (Heidegger 1998, 185). Emphasis is his.

[2]Galileo famously railed against Aristotelian science, to the great chagrin of his university professors, and later, when he was a member of the mathematics faculty at Padua and then at Venice, to the consternation of his colleagues, later still, to the contempt of the church. But, Galileo was not doing Aristotle a disservice. In fact, Galileo was a better reader of Aristotle than were Aristotle's greatest supporters, which is to say that Galileo appreciated Aristotle for what he tried to do but put in the work to prove some of Aristotle's most well-known but incorrect conclusions erroneous. In *Letters on Sunspots*, Galileo in fact defends Aristotle against his poor readers, suggesting that while Aristotelians of his day are satisfied with defending what is false, Galileo imagines Aristotle would have been interested in discovering the truth (Galileo 1957, 118 and 142).

[3]Newton's *Principia* was published in 1687.

along by Copernicus, Descartes, Kepler, and Galileo—it became seemingly antiquated.[4]

The question as to whether Aristotle's *Physics* is a work of philosophy or a work of science has not been, and really cannot be, properly addressed. But, whereas scientists largely consider older works of science to be passé, philosophers still glean important lessons from ancient texts.[5] In service primarily to the latter, but also to the former to the extent that I think Aristotle still has something to teach us about how and why to pursue scientific questions, I have undertaken the present work. In what follows, I offer an interpretation of the first half of Aristotle's *Physics*, his most general work in natural philosophy, where he famously sets out to uncover the simplest elements (*stoikeia*), primary conditions (*aitia*), and first principles (*archai*) of nature (184a9–16) and defines nature (*phusis*), motion, *kinâsis*, the infinite (*apeiron*), place (*topos*), void (*kenon*), and time (*chrónos*).[6] The interpretation and my arguments that guide it are with an eye to developing a coherent way to understand Aristotle's approach to knowing and defining these notoriously difficult concepts in physics. Specifically, my target is Aristotle's Treatise on Time (*Physics* iv 10–14),[7] which comes at the end of *Physics* iv.[8] I

---

[4]Maudlin (Maudlin 2012, 4) explains: "Abandoning Aristotle's spherical universe entails abandoning his basic physical principles and rethinking the form that the laws of physics can take. This task was undertaken by Sir Isaac Newton."

[5]Even in the philosophical community, however, interest in Aristotle's philosophy of nature is relatively recent in the modern and post-modern periods, having received little attention until the mid-late twentieth century (see Couloubaritsis 1997, 2-4). At this time, a series of commentaries emerged (Mansion (1913) 1946, Solmsen, 1960, Wieland (1962), 1972, Heidegger, 1967), spearheaded by Sir David Ross's 1936 monograph, now considered the standard work on the Physics.

[6]Aristotle in fact redefines nature in the *Physics*, seemingly unseating what has been argued to be the Platonic conflation of nature with being (see Grant 2006) and the various Pre-Socratic identifications with nature as matter, an organizing principle, or both.

[7]Recent scholarship in ancient philosophy has seen a renewed interest in Aristotle's Treatment of Time in the *Physics* (*Physics* iv 10-14). And, it is the account of time that is again calling attention to other parts of the work. This is to say that recent contributions to the growing literature on Aristotle's Treatise on Time (*Physics* iv 10-14) have focused on the importance of considering this section in light of arguments made in other sections of the work (Couloubaritsis 1997, Coope 2004, Roark 2011, de Moor 2012). In *Time for Aristotle's* Physics *IV. 10-14*, Ursula Coope argues that Aristotle's account of change, and in particular, Aristotle's argument that change is divisible, is required to understand his account of time (Coope 2004, 5–9). Similarly, in *Aristotle on Time: A Study of the* Physics Tony Roark attempts an "hylomorphic interpretation" of Aristotle's account of time (Roark 2011, 1–8). He argues that time is "a combination of matter (*hule*) and form, or shape (*morphe*)" (Roark 2011, 1) analogous in structure to the natural substantial beings Aristotle investigates in the *Physics*. Most recently, following Lambros Couloubaritsis, in *Aristote et la Question du Temps* (de Moor 2012, 133), Mieke de Moor argues, "the question of time in the *Physics* is not a question independent of, but closely linked with, the framework in which it is posed." Translation is my own. De Moor likewise provides a near exhaustive history of *chrónos* in the time preceding Aristotle.

[8]Simplicius, among other early commentators, believed the eight books of the *Physics* to be an amalgamation of two original texts, one on nature and one on motion (*kinêsis*). On one account of

focus on chaps. 10 and 11, where I locate Aristotle's analytic of time. I explain the analytic in terms of Aristotle's foregoing arguments in the *Physics*, i.e., in terms of its context.

For Aristotle, the nature of natural beings is an inner principle of motion and rest, motion is an actuality (*entelecheia*) of the potentiality (*dunamis*) for the acquisition by a substantial being of an accidental form, the infinite is never actual and always a potential for continuous division or addition, place is a boundary occupied by a body, and void either does not exist in any modality, or—on some readings—it exists in potentiality as rarefaction. Aristotle's definition of time, "number of motion in respect of 'before' and 'after'" *ἀριθμὸς κινήσεως κατὰ τὸ πρότερον καὶ ὕστερον* (219b1), thus depends on (1) what he means by number, (2) what he means by motion, and (3) what he means by before and after. Further, what Aristotle means by motion turns out to require one to understand what he means by potentiality, actuality, and nature more generally. Thus, the nature of time for Aristotle is inextricable from his wider study of nature, including not only arguments Aristotle makes in the *Physics*, but also to include points taken from the *Metaphysics*, the *Nicomachean Ethics*, much of the Organon: the *Categories*, *De interpetatione*, *Analytica Posteriora*, *Topics*, and works on more narrow topics from natural philosophy: *De anima*, *De sensu et sensibilibus*, *De memoria et reminiscentia*, *Historia Animalium*, *De motu animalium*, *De generatione animalium*, *De partibus animalium*.

As some have before me, I emphasize the role of potentiality in Aristotle's general account of nature in the *Physics*,[9] but I do so by calling attention to the key

_____

(Footnote 8 continued)
this division, it is possible that the time section was in fact the end of the work on nature (See Ross 1936 on the possibility that the work concerning motion began with book five). Possible support for this proposal is that Aristotle's allusions to time in *Physics* v–viii align with only one of the two types of time, i.e., infinite time, he names at the beginning of his Treatise on Time (218a1); whereas, what I designate as his analytic of time, in his Treatise on Time, focuses on another type of time, i.e., time taken. It is for this reason that the present work concludes its examination at the end of *Physics* iv.

[9]In *Time, Creation, and Continuum: Theories in Antiquity and the Early Middle Ages*, Richard Sorabji (Sorabji 1983, 90) acknowledges interpreters long have been confused by Aristotle's claim that the existence of time requires countability and thus someone to count (Cf. *Physics* iv 14, 223a25). He believes that Aristotle is mistaken about this, but reconciles the apparent confusion pointing to Aristotle's intentions and preoccupations. While I do not agree with Sorabji when he says that Aristotle's definition of time is wrong, I do agree with his assessment that readers of Aristotle's Treatise on Time need to, "turn our attention away from time to the notion of *possibility* and to such related modal motions as countability" (his emphasis). Sorabji's point turns us to one of Aristotle's primary interests in his natural philosophy—potentiality. I add to Sorabji's critique and suggest that the confusion he cites on the part of commentators has led to a general misunderstanding of Aristotle's method and goals in the *Physics*. Namely, it is sometimes claimed that Aristotle's *Physics* in general, and the Treatise on Time in particular, is a confused inquiry—part metaphysics and part epistemology. On such a reading, Aristotle can be understood in such a way that it looks like he thinks time is actually existing, i.e., "is real," with or without the potentiality for perception and intellection in the world, i.e., without "someone to count." I consider this reading to take Aristotle's Treatise on Time out of its intended context and to impose on his work a

role of the interplay between potentiality and actuality in Aristotle's treatments of natural beings and the physical concepts. Thus, I argue that Aristotle's conclusions about time are analogous to Aristotle's conclusions about the infinite, place, and void, i.e., that they are derived from his theory of nature generally.

To the extent that time is derived from nature, it is not a self-subsistent being *qua* itself. Instead, it is always only a potential being (insofar as *kinêsis* can exist independently of its apprehension) unless it is apprehended by one or more faculties of cognition, e.g., perception, phantasia, intellection. In this reading, time for Aristotle emerges as an actualized potentiality, which exists as actual as a result of an interaction between two or more parts of nature, viz., self-subsistent natural beings constantly changing modal status as part of their very being.[10] The parts of nature that can relate in such a way as to bring about actual time require conditions for the actualization. After I work through the proposal that the Treatise on Time is a parallel account to Aristotle's treatments of nature, motion, the infinite, place, and void, therefore, I take seriously what Aristotle says in *Physics* iv 14 regarding the role of soul in the existence of time. Finally, I complement my reading of Aristotle with supporting evidence from contemporary animal science to show that even non-human animals on Aristotle's account have some sense of time. In sum, time for Aristotle is never an *a priori* or fixed presence; it is not a container, a continuum, or a copy of eternity. What we know of time—indeed, what we make of it, is the same thing as what time actually is.[11]

(Footnote 9 continued)

bifurcation between a study of reality and a study of knowledge, i.e., a difference between *what is* and *what we can know*, which does not apply in Aristotle's time. This is not to say that Aristotle does not allow that there are a very many things about which we do not know, only that these things are outside the scope of his investigations in natural philosophy. Here, he allows that nature is self-evident and his investigations follow from this initial premise. Thus, that after which he inquires, and so too, that which he seeks to know and that about which he comes to know, is necessarily derivative of that to which he has perceptual and intellectual access. Sorabji looks at Aristotle's use of *aesthesis* in four works to justify the claim that Aristotle is concerned with possibility in the *Physics* generally, so too in the Treatise on Time.

[10]The question as to whether time is "real" or "unreal" for Aristotle is a question about whether or not time is an existing self-subsistent being (οὐσίαν αὐτὸ ὄν) for Aristotle. Demetra Sfendoni-Mentzou argues in her paper, "Is Time Real for Aristotle?" (2012), that time is indeed real for Aristotle. Inwood (in Judson 1991, 177) also holds this view. However, Inwood limits his study to an investigation of the puzzles of time in *Physics* iv 10 and does not treat Aristotle's provocative discussion about time and the soul in *Physics* iv 14.

[11]It has been popular for readers of Aristotle's Treatise on Time to label him with a certain epistemological orientation. Namely, Aristotle has been called an "idealist" as a response to misunderstanding the role of *nous*, or the requisite "counter," in his Treatise on Time. This characterization of Aristotle thus sets up another polarization of his position, against the typical portrayal of Aristotle as "realist," but—more importantly—reducing Aristotle's dynamic work to anachronistic and overly simplified terms. Aristotle holds nature to be self-evident; but, he does not understand time to be a natural being qua itself. The idea that actual time—time taken—requires a taker is not to say that Aristotle has switched metaphysical camps. Rather, Aristotle is arguing for a time concept, which develops from the way natural beings interact—the way they are in the world and the way they *take in* the being of others.

In the end, despite various scientific inaccuracies the *Physics* might present as true, we have an important lesson to recover from Aristotle's philosophy of nature and, specifically, from his nuanced Treatise on Time—an inquiry into nature and its attributes is always also an investigation of ourselves, not only of our own place in the general context of nature, but also what we contribute to it to the extent that we are always already interacting with the context we endeavor to understand. For Aristotle, time results from particular kinds of interactions among different beings in nature. It is not only this result, but also the modality of this interaction to which this monograph endeavors to respond.

Chapter 1
Time in Context

Aristotle was not a philosopher concerned with time—in questions about time or in delimiting the being of time.[1] Instead, Aristotle was a natural scientist interested in the being of natural things, whose ways of being demand a discussion of time.[2]

Thinking backward, time (*chrónos*) is an attribute of motion or change, (*kinêsis*), for Aristotle (219b1–2). Aristotle's interest in time and in that which is supposedly required in order that there be motion, e.g., infinity, place, and void (200b12), come from his interest in motion itself, and motion only because he sought in the *Physics* comprehensive understanding of the fundamental principles of natural beings. His analysis of time, then, comes from his interest in the study of natural being. If we can read Aristotle's account of time in this way, then we will be able to see that his account of time cannot be an account of time *qua* time.

In this chapter, I begin with a careful reading of *Physics* i 1, emphasizing the importance of acquiring a sense for the scope, goals, and method of Aristotle's project as preparatory to reading subsequent arguments in the work. In particular, I propose that, contrary to the typical polarized readings of this section, Aristotle's method in the *Physics* is necessarily a combination of dialectic, beginning with *endoxa*, and demonstration, beginning from experience.[3] I then continue my reading of book 1, highlighting the potentiality (*dunamis*)/actuality (*entelecheia*) distinction in Aristotle's account of natural change, *gignomenon* and *kinêsis*, in

---

[1]Aquinas (1961, 2) seems to be making a similar claim, though not explicitly, when he writes that natural science "deals with those things which depend on matter not only for their own existence, but also for their definition." According to Aquinas's prior argument, subjects like number, magnitude, and figure depend on matter for their existence but not for their definition. By deduction, since time is a number for Aristotle, it would not be a proper subject qua itself for natural science.

[2]Roark calls Aristotle a philosophical optimist because Aristotle is willing to define time (Roark 2011, 11). Without intending to take anything away from Aristotle, I would disagree with Roark on this point. Aristotle is in fact not defining time as the abstract concept we have come to know in contemporary discourse. Aristotle's naturalist account of time was not over-reaching; he had no motive for saying any more about time than his project warranted.

[3]The sense in which I intend demonstration here has to do with that which utilizes experience (*apodeixis*) instead of demonstration using only pure thinking, e.g., a demonstration by way of syllogism based on definition.

© The Author(s) 2015
C.C. Harry, *Chronos in Aristotle's Physics*,
SpringerBriefs in Philosophy, DOI 10.1007/978-3-319-17834-9_1

substantial natural beings. I suggest that this distinction is quite important for Aristotle's conclusions about the principle of natural beings in *Physics* ii (192b13–22) and that understanding these conclusions entails that the potentiality/actuality distinction is preparatory to a full reading of the definition of motion Aristotle advances in *Physics* iii 1 (201a10–11) and to his subsequent explanations of the infinite (iii 4–8), place (iv 1–5), void (iv 6–9), and especially time in the *Physics*. Namely, Aristotle is going to argue that the natural scientist must think radically differently about the ontological status of motion and the terms of motion, i.e., they are not self-subsistent natural beings. The following chapter will thus be divided into the following subsections: (1) *Physics*: Scope, Access, Goals, and Method; (2) Nature and its *Archai*; (3) The Role of *Kinêsis* in Nature; (4) From *Kinêsis* to *Chrónos*.

1.1 *Physics*: Scope, Access, Goals, and Method

Aristotle's *Physics* is a book about nature (φύσις, *phusis*). It is an inquiry into nature itself, but this means an inquiry into the objects of nature in so far as these objects seem to reveal what nature is, i.e., the principle of their motion.

The first line of *Physics* i 1 introduces the scope of the project: "When the objects of any inquiry (πάσας τὰς μεθόδους), in any department, have principles (ἀρχαὶ), causes (αἴτια), or elements (στοιχεῖα), it is through acquaintance with these that knowledge and understanding is attained" (184a10–12). Aristotle's goal is not a subjective objective to overcome skepticism; rather, he is searching for comprehensive understanding.[4] Such comprehensive understanding is by way of getting to know the principles, causes, and elements of the subject of inquiry. If nature generally, i.e., nature *qua* nature, is Aristotle's subject, then such an acquaintance is implausible. A project with such a large scope seems rather Platonic.[5] Instead, his subject must be the nature of the various natural beings. The objects of nature can be investigated, probed, and analyzed for the grasp of the natural principle involved in each kind of natural being.

Aristotle continues, announcing that the first task in the present inquiry, which he names a science of nature (φύσεως ἐπιστήμης), will be to "determine what relates to its principles" (διορίσασθαι πρῶτον τὰ περὶ τὰς ἀρχάς) (184a15–16). The question here at the beginning of the *Physics* is not only about the scope of inquiry, but also about access to the subject. To know the principles, one must make a determination about what concerns them. Aristotle's point here is subtle, but if acquaintance with the principles, causes, and elements of natural beings is the

---

[4]According to Aquinas (1961), *Physics* i 1 is a preface to the rest of the work.

[5]Elsewhere, Aristotle explicitly distances himself from Platonic-style natural philosophy. See 203a16 and *Meta* 1001a12 on this point. As Ross points out, Pythagoreans and Plato were "thought of as being a priori theorists rather than genuine students of nature" (Ross 1936, 545).

scope, and if comprehensive understanding is the goal, the natural scientist must have both (1) the potentiality for perception of that which concerns the principles of nature, and (2) a faculty of judgment in order to acquire knowledge of them.

Scope and access must be buttressed by a clear method. Aristotle famously outlines a method, but what exactly he intended to convey is disputed. He will gather knowledge of these principles moving "from what is better known to us to what is better known by nature" (πέφυκε δὲ ἐκ τῶν γνωριμωτέρων ἡμῖν ἡ ὁδὸς καὶ σαφεστέρων ἐπὶ τὰ σαφέστερα τῇ φύσει καὶ γνωριμώτερα) (184a16). On one reading, "what is better known to us" may mean *endoxa*, or those ideas that are commonly held. On this reading, the method Aristotle proposes is dialectic (see also *Topics* viii 5), and the ideas about nature whence he would be starting would have been those advanced by his predecessors, e.g., Plato, the atomists, the sophists, and the Eleatics.[6] On another reading, "what is better known to us" is that which is more readily perceived.[7] The discrepancy over interpreting this passage is pertinent because the way one understands Aristotle's method grounds the way one reads the rest of the work. When there has been relatively little attention to *Physics* i 1,[8] and instead an interest only in later books, this essential topic is left unacknowledged. Is Aristotle primarily a philosopher concerned with evaluating the ideas of his pre-decessors,[9] or is he a natural scientist, looking for clear demonstration of the nature of natural beings? Having left this question unanswered, and indeed unasked, some modern readers have supposed that Aristotle's *Physics* is inconsistent in method— changing between metaphysics of nature and an epistemology.

As was aforementioned, the potential for perception and judgment are both pertinent to achieving the goals Aristotle has laid out. If sense perception is indeed integral to the project, he is going to acquire knowledge of the nature of natural beings at least in part by way of demonstration. Consider, for example, Aristotle's general account of perception and knowledge in *De anima* ii 5 (417b17–28) where

---

[6]See Hussey (1983), Owen (1986), and Irwin (1988), all proponents of this view. Hussey explains that "*endoxa*" need not mean exclusively ideas commonly held: "His method is the method of dialectic, by which (in theory at least) the philosophical inquirer started from the accumulated material of common-sense intuitions, previous opinions of philosophers, and observed facts relevant to the subject, and ascended by a process of rational criticism and generalization to the correct account of the subject, which would usually be enshrined in a definition of the central term" (ix).

[7]See Bolton (in Judson 2003), for a very well argued take on this view. For Bolton, *Physics* i 1 is a parallel account to *Posterior Analytics* ii 19. According to these accounts, "the conclusion of our reasoning and our inquiry gives us a principle which *explains* (and gives us a firm delineation of) the perceptible phenomena which we use to reach it. But no rule of general dialectic or of any type of dialectic which Aristotle discusses is designed to guarantee conclusions of this sort. So if dialectic does reach conclusions of this type it is accidental and not due to the method of dialectic itself" (13).

[8]This is true of modern commentators, but as we will shortly see, Aquinas did comment at length on the importance of understanding Aristotle's method.

[9]In an effort to laud the development of the scientific method in the sixteenth century, modern physicists oversimplify Aristotle's project stating that he had no method of demonstration and relied entirely on dialectic to obtain conclusions about nature.

he asserts that scientists must have access to external perceptibles in order to acquire knowledge of things generally. Because there is no one place to start an inquiry into nature when the subjects of the inquiry exist external to the scientist,[10] with what the scientist begins her inquiry—with *endoxa* or confused perceptions— should not be a disjunction. Instead, it should be a conjunction; the two are inextricable in natural science.

Aristotle provides further detail about his method. These paragraphs are notoriously difficult to interpret because Aristotle uses the terms universal and particular equivocally. Since he has just insisted that we begin with things more knowable to us, he then qualifies this point saying: "For the same things are not knowable relatively to us and knowable without qualification" (οὐ γὰρ ταὐτὰ ἡμῖν τε γνώριμα καὶ ἁπλῶς) (184a18).[11] Read one way, "knowable relative to us" refers to the natural scientist's access to external perceptibles. The scientist obtains knowledge about universals, meaning genera, by experience with different kinds. The external perceptibles we encounter through sense perception are the particular instantiations of universals whence we acquire general knowledge.[12] Read another way, "knowable relative to us" refers to the ideas commonly held prior to this investigation. How we read Aristotle's proposition here affects how we understand his subsequent conclusion. If he formerly intended to say that we garner knowledge of genera by way of experience with kinds,[13] his conclusion that his method will: "advance from what is obscure by nature, but clearer to us (ἐκ τῶν ἀσαφεστέρων μὲν τῇ φύσει ἡμῖν δὲ σαφεστέρων), towards what is more clear and more knowable by nature (ἐπὶ τὰ σαφέστερα τῇ φύσει καὶ γνωριμώτερα)" (184a19–21), means that the scientist will begin with the individual natural beings whose general nature she wants to know better. If, however, Aristotle meant that we begin with ideas commonly held, then the conclusion announced about his method entails working from the ideas of predecessors. In this case, the method does not clearly involve scientific observation or investigation.

---

[10]Being a self-subsistent natural being herself, the natural scientist could be a subject to herself, but this is not likely what Aristotle had in mind.

[11]Aquinas (1961) explains the difference between knowable to us and knowable by nature/without qualification with an appeal to the fact that humans begin from potency and from a point of view of science and nature's *telos* is something to be learned. I disagree with Aquinas on this point. It seems a Christianized reading of Aristotle here, separating humans from the natural order, and especially from God. Aquinas's conclusions may be the first source of commentary reading Aristotle as part epistemology (studying what we can know) and part metaphysics (studying what is).

[12]This passage could be misread to suggest that Aristotle is differentiating what is possible for us to know, or an epistemology, and what is, a metaphysics. This is not the case. He is not insinuating that there is, to use a Kantian term, a noumenal aspect of nature that escapes our grasp. Instead, he is introducing his understanding of knowledge acquisition as (1) perception of particular instantiation of a genus (2) knowledge of universal (genera) from experience with the particular.

[13]Aquinas writes against Ibn Rushd's claim that Aristotle meant, "composed" when he tells of confused first perceptions. Aquinas disagrees with this reading because, as he points out, genera are not *composed* of species.

Aristotle adds here that we will get at the first principles by starting with "rather confused masses" (τὰ συγκεχυμένα μᾶλλον) (184a22), or, as Ross renders it, "the confused data we start with" (Ross 1936, 15), and by subsequent analysis, we will achieve the anticipated elements and principles (ὕστερον δ' ἐκ τούτων γίγνεται γνώριμα τὰ στοιχεῖα καὶ αἱ ἀρχαὶ διαιροῦσι ταῦτα) (184a22). Again, one could understand "confused masses" to mean either *endoxa* or first perceptions and "analysis" to refer to dialectic or demonstration. But, what Aristotle writes next points us to the latter explanation. He clarifies that we first come to know universals (καθόλου) and then particulars. Of course, universal is used here in a sense different from the one just discussed, i.e., universal as genera. Here, universal refers to a whole, and that is contrasted with particular, meaning part of the whole. These universals, or wholes, are better known to sense than particulars (διὸ ἐκ τῶν καθόλου ἐπὶ τὰ καθ' ἕκαστα δεῖ προϊέναι· τὸ γὰρ ὅλον κατὰ τὴν αἴσθησιν γνωριμώτερον, τὸ δὲ καθόλου ὅλον τί ἐστι· πολλὰ γὰρ περιλαμβάνει ὡς μέρη τὸ καθόλου) (184a23). What is better known to us, then, is what is better perceived by us; and, wholes are perceived before parts. To illustrate this point, Aristotle uses the example of the child first associating the name "father" with all men and then later determining that only one of those men is his own father. Similarly, a child learns "dog," "cat," "chair," "table," and only over time learns what distinguishes one instance of these universals from another. The demonstrative, "that cat" or "this chair" develops into a more specific identification: "the Adirondack Chair my grandfather owned" or "the black Labrador Retriever named Wolfgang." Even as adults, when we first experience objects, we do not immediately distinguish among various instantiations of them. We see a starling for the first time, and subsequent sightings of different starlings are indistinguishable from the first.

For Aristotle, this fact is explained analogously by the difference between names and definitions. Names, he clarifies, do not differentiate among wholes themselves; whereas, definitions differentiate wholes into particulars, i.e., a definition associates a species with a genus and also differentiates the species from other kinds of the genus. Therefore, our perceptions do not immediately differentiate among that which is of the same kind. Our analytical abilities organize and categorize. But, this is not just a point about language. For Aristotle, the confused perceptions we first notice are compounds (184a23).[14] We see the whole and the parts together as one, and then later we perceive the detail. With regard to the science of nature, we will come to know nature by intentionally analyzing our immediate perceptions in order to learn about particulars not immediately clear to us. Following the logic here, then, in order to know the first principles of nature, we will analyze our perceptions of natural beings.

---

[14]As Aquinas (1961, 5–6) points out, that Aristotle formerly said confused and not compound is significant, as he is using universal equivocally: integral sensible, universal intelligible, universal sensible.

This is the end of the first chapter of book 1 and so too of the discussion of scope, access, goals, and method. Aristotle's method in the *Physics* is more complex than the one either of the two usual positions attribute to him.[15] Further, holding him to one or the other is likely a false dichotomy. It seems clear that Aristotle was a natural scientist setting out to demonstrate the nature of natural beings in so far as this is consistent with his aforementioned scope, goals, and requisite access to the subject of inquiry. But, it does not seem likely that he would not have found it necessary to engage with the commonly held ideas at his time. In fact, he had an onus to contend with the ideas of his predecessors. He was arguably the first natural scientist because he was not simply speculating about nature. But, in being the first, the scope of his project took on an implied second dimension, i.e., he needed to show that his new way of explaining nature was valid. This is not to say that Aristotle would have denied the importance of thinking through ideas logically, but just that he set himself apart from his predecessors because he saw the parallel importance of justifying scientific conclusions with demonstrable evidence based on experience. Drawing attention to this debate in *Physics* i 1 and attempting a non-polarized reading of it establishes certain expectations with regard to Aristotle's arguments in later books. It prepares us not only for Aristotle's complementary use of dialectic and demonstration, but also prevents us from reading these later arguments in modern terms—as "metaphysics" at one turn and then as "epistemology" at another.

Aristotle now begins his explorations to uncover the principles of the nature of natural beings. Already in the next chapter we will see him switching between demonstrations from experience, beginning with confused perceptions, and dialectic, beginning with *endoxa*. Having now suggested that Aristotle's method is not strictly in one vein or the other, I will point out his movement between them in my following discussion of his arguments.

1.2 Nature and Its *Archai*

After establishing the scope, goals, access, and method for natural science, Aristotle sets out to know the principles of nature of natural beings. By way of a discussion of these principles, Aristotle works up to a discussion of potentiality and actuality in his account of natural change, *gignomenon* and *kinêsis*, in substantial natural beings. The role potentiality and actuality play in Aristotle's observations of and conclusions about the nature of natural beings help us to understand his subsequent discussions of the infinite, place, void, and especially time.

---

[15] Aquinas (1961, 3) believes that already in the first line of the work we see Aristotle wedging a difference between understanding and science, disclosing the importance of both definitions and demonstrations to natural science.

He begins in *Physics* i 2 with a significant assumption; one that has been supposed prior to his announcement in *Physics* i 1 that his goal is to determine what relates to the principles of the nature of natural beings[16]; namely, that there is in fact a principle of natural science (184b14–15). Determining whether the principle is one or more will not be a matter of analyzing perception, but of weighing the logic of various *endoxa*.[17] Either, the principle is one and unchanging, a view attributed to Parmenides and Melissus, or the principles are more than one and subject to change, various examples of which are advanced by the natural philosophers. If the principles are more than one in number, then they are either limited in number or unlimited in number, and they are either all the same or different in kind. These are not propositions based on experience. Rather, they are based on logical possibilities (see 184b26–185a10). But, this is not to suggest that Aristotle will conduct the entire examination as a conversation with the *endoxa*. Rather, these are the positions with which he must contend before he can use perceptibles to try and establish a different view.

In fact, Aristotle proceeds to belie the *endoxa*, arguing that the scope of his project requires a method of investigation *different from* the one used by his predecessors. He reemphasizes here that the subject of his inquiry is natural beings. These beings are known by perception and understood by analysis and discrimination: "We, on the other hand, must take for granted that the things that exist by nature are, either all or some of them, in motion—which is indeed made plain by induction" (Ἡμῖν δ' ὑποκείσθω τὰ φύσει ἢ πάντα ἢ ἔνια κινούμενα εἶναι· δῆλον δ' ἐκ τῆς ἐπαγωγῆς.) (185a14). Aristotle's analytic will begin with the basic fact of that which is evident to perception, side stepping purely speculative argument. And, it is perception that allows for Aristotelian natural science, not an a priori science. If he cannot debate with the *endoxa* on their own terms, he has to rely instead on an alternative method.

According to Ross (1936, 487), in our current chapter divisions, it is in *Physics* i 5 that we get the beginning of Aristotle's analytic of the first principles.[18] Let us

---

[16]Aristotle made a quick switch from the hypothetical, announcing what is possible, to the actual, announcing how he will actually proceed, at the start of this work: "when the objects of an inquiry...have principles...it is through acquaintance with them that knowledge and understanding is attained" to "in the science of nature...our first task will be to try to determine what relates to its principles" (184a10–16).

[17]Bolton (in Judson 2003) argues that Aristotle's use of *endoxa* here does not undermine his promised engagement with perceptibles in *Physics* i 1. I agree with his conclusion that engaging with the *endoxa* in *Physics* i 2 is complementary to his otherwise demonstrative methodology. Specifically, Bolton supposes Aristotle to be exercising the following point: "the natural scientist cannot use a scientific, that is a demonstrative, argument to refute someone who denies that the natural world of changing things exists. In natural science it is an indemonstrable first principle that the natural world of changing things exists...one can only refute this denial dialectically, or peirastically" (15).

[18]I will continue to highlight that Aristotle reserves his own thoughts on the various topics of natural science until after having contended with the *endoxa*. We will see this pattern in his subsequent discussions of nature, motion, the infinite, place, void, and time.

trace Aristotle's analytic beginning at 188a19. Aristotle defends the point that the principles must be at least two. In so doing, he concludes that contraries (τἀναντία) are the principles, re-invoking the *endoxa* to show that both Parmenides and Melissus, and the natural philosophers, allow for this to be the case.[19] It is in this chapter that we see Aristotle really moving between his proposed method of progressing from first perceptions or wholes to discriminating parts and weighing *endoxa*. He begins by showing that even the monists must agree that contraries are principles because they posit fire and earth. Aristotle likens this to a commitment to principles of hot and cold. Likewise, he cites a belief in the dense and rare and then the atomist belief in the full and empty (188a20–25). After establishing that all agree contraries are principles, Aristotle clarifies that the principles, or primary contraries, must be foundational, giving rise to all else. Aristotle then announces what will be a turn to logical considerations. We will determine the need for primary contraries not by way of considering what is the case in our perceptions, but by thinking through what would or would not make sense to conclude about the character of contraries. Aristotle's take on contraries here will open up the possibility for thinking non-being or potentiality in nature.

Aristotle uses examples from experience. He has us think about the qualities of paleness and knowing music and asks rhetorically how it could be possible that the latter come from the former (188a36). He uses what his reader would readily admit from experience to demonstrate the difference between logical contraries and logical complements. White does not come from any not white, he writes, and what he means is that white does not simply come from not white. For example, a wise person is not white, but the person is not not-white, which is reserved for a color other than white, especially the contrary, black. Aristotle explains further that the "white" can come from an intermediary of it and its contrary, in this case black. But, white does not come from a complement, i.e., a non-color in this case, something merely non-white. He continues, but it is unclear whether he is speaking from experience, or if he is explaining in terms of what makes sense. Nothing changes into something categorically different, except by chance. White does not turn into wise or properly trained; it changes into its contrary or something between the two (188b4–8). Rather, the thing—such as a pale person—may become properly trained or wise.

But, this is not the only sense in which one contrary becomes another. When we are talking about a change from white to not white, we convey a change in predication of a substantial being. We can also talk about general not being changing into being. Ross (1936, 489) calls these contraries "the thing produced," i.e., fitted together (ἡρμοσμένον) and "that from which it is produced," (ἀνάρμοστον). Aquinas explains that these contraries are called primary contraries because in order for the latter to be principles, they would require principles themselves

---

[19]See Ross (1936, 487) for evidence that Aristotle probably mistook what was said in Parmenides's poem for Parmenides's own views.

(Aquinas 1961, 48).[20] To illustrate this point, Aristotle discusses compound being that comes to be from a state in which it was not. A house is built from materials that when grouped together as they lie on the ground are "not a house." But, these materials are quite different from a bunch of chemistry equipment lying on the ground, which also constitute things that are not a house, but now in a complementary sense. They have no possibility of ever being a house. For Aristotle:

> It does not matter whether we take attunement, order, or composition for our illustration; the principle is obviously the same in all, and in fact applies equally to the production of a house, a statue, or anything else. A house comes from certain things in a certain state of separation instead of conjunction, a statue (or any other thing that has been shaped) from shapelessness—each of these objects being partly order and partly composition (188b15–20).[21]

The two senses of contrary here, and thus the two senses of change that have emerged, will foreshadow Aristotle's later discussion of accidental formal change and substantial change.[22] On the one hand, we see him developing a theory about accidental formal change. When something, e.g., the pale person, is white and undergoes qualitative change, it becomes not white, meaning it moves either closer to its contrary, black, or it moves completely to black. There is a principle of accidental change, but the substance is already in place. For example, the person is now untrained and then properly trained. On the other hand, above we see Aristotle describing a principle of moving between contraries where the contraries signal generation and corruption, instead of accidental formal change. Here, we see that the coming into being of a complex substance requires both the material components and then imposition of the form. And, certain forms require certain materials, or the substance cannot come into being. Aristotle concludes that everything coming to be naturally is either a contrary or a product of contraries (188b24–26). In all cases, as Aristotle works to show, contraries are the source of change.

As Aristotle admits, his predecessors would not have disagreed that the *archai* are contraries. Yet, while most of his predecessors have asserted what they believed to be principles as contraries, what exactly the contraries end up being varies widely (188b37–189a2). Again, we see a distinction being made between confused perception, here called "the order of sense" (κατὰ τὴν αἴσθησιν) and logical reasoning, or, "the order of explanation" (κατὰ τὸν λόγον). Aristotle assigns the coming to know by way of explanation to the universal and the coming to know of the particular to the order of sense: "The universal is more knowable in the order of

---

[20]See also Aristotle on this point at 189a30–35.

[21]διαφέρει δ' οὐθὲν ἐπὶ ἁρμονίας εἰπεῖν ἢ τάξεως ἢ συνθέσεως· φανερὸν γὰρ ὅτι ὁ αὐτὸς λόγος. ἀλλὰ μὴν καὶ οἰκία καὶ ἀνδριὰς καὶ ὁτιοῦν ἄλλο γίγνεται ὁμοίως· ἥ τε γὰρ οἰκία γίγνεται ἐκ τοῦ μὴ συγκεῖσθαι ἀλλὰ διῃρῆσθαι ταδὶ ὡδί, καὶ ὁ ἀνδριὰς καὶ τῶν ἐσχηματισμένων τι ἐξ ἀσχημοσύνης· καὶ ἕκαστον τούτων τὰ μὲν τάξις, τὰ δὲ σύνθεσίς τίς ἐστιν.

[22]This distinction may indicate Aristotle's early thoughts on the difference he will make in *Physics* v between *kinêsis*, usually rendered "motion" but generally meaning accidental or predicative change, and *metabole*, which includes *kinêsis* but also substantial change, i.e., generation and corruption.

explanation, the particular in the order of sense: for explanation has to do with the universal, sense with the particular" (189a5–9).[23] Of course, Aristotle is using universal and particular equivocally again. Here, he uses the same sense of universal that we saw in his general account of perception and knowledge from *De anima* ii 5, i.e., universal as genera. We come to know universal as genera through experience with particulars, or kinds; we distinguish the form from the material. This sense of universal is obviously different from the meaning Aristotle intended when he said in i 1 that the child first calls all men father. In this case, we only know the confused whole until we discriminate it into parts.

When Aristotle is talking about *endoxa*, he differentiates among those predecessors who named the principles as contraries more knowable by sense and those that are more knowable by explanation. Thus, Aristotle himself is separating theories that have arrived at principles by way of different methods. The theories that have unobservable unified concepts as principles—Plato's great and small, for example, or Empedocles's love and strife—have based their physics on universal principles that allow us to account for the whole. The theories that have basic observable phenomena as principles—Anaximenes's account of the dense and the rare, for example—allow us to conjecture something about the whole based on experience of whatever is in front of our senses. For Aristotle, these accounts are analogous; both are providing a scheme for the order of things in nature. But the nature of the accounts marks a difference among them; Aristotle sees a value and a necessity in contending with physics based on theory, his analytical work is based on what can be perceived.

Aristotle concludes this discussion, clear that the *archai* are contraries. His view has been justified using appeals to experience as well as rational explanation. In addition, he has corroborated his claim with the wide range of *endoxa* from his time. So, he begins a new discussion with a follow-up question. Even if the principles are contraries, are they two or more than two (πότερον δύο ἢ τρεῖς ἢ πλείους εἰσίν) (189a11)? He proceeds to argue that they are more than two, but finite in number.

This portion of Aristotle's argument is quite important because it is here that he introduces both the idea of an unchanging element of nature, ὑποκείμενον, which he will develop later when he defines nature as a principle of motion, and the issue of multiple causes, which Aristotle will famously develop in *Physics* ii. As Aristotle begins to suggest that the principles are more than two, yet finite, he uses examples where it is unclear if he is speaking from experience or from a point of view of logical necessity. He says, for example that, "it is difficult to see how either density should be of such a nature as to act in any way on rarity or rarity on density" (189a22–23). He then supports this view invoking a second example from the *endoxa*: neither can love and strife gather one another up and make something out

---

[23]τὸ μὲν γὰρ καθόλου κατὰ τὸν λόγον γνώριμον, τὸ δὲ καθ᾽ ἕκαστον κατὰ τὴν αἴσθησιν ὁ μὲν γὰρ λόγος τοῦ καθόλου, ἡ δ᾽ αἴσθησις τοῦ κατὰ μέρος.

of each other. It seems apparent that there must be some "third thing" (ὑποτιθέναι τι τρίτον) (189a25). Recall that Aristotle has been discussing contraries in two different ways: as true contraries, e.g., white and black, and as primary contraries, e.g., what Ross calls "the thing produced" and "that from which it is produced." Thus, this third thing is going to mean something different for each of these pairs of contraries.

Aristotle wants to posit a third thing, similar in the vein of Pre-Socratic philosophers of nature (189b1–2). More than three would be inefficient because we would end up with more than one contrary, and we would thus require an intermediary for both. But, he will not do that here. First, he provides his own account, retracing his arguments from *endoxa* and effectually supplanting them with arguments from confused perceptions. He will now turn to the "natural order of inquiry," i.e., from common characteristics to particular cases (ἔστι γὰρ κατὰ φύσιν τὰ κοινὰ πρῶτον εἰπόντας οὕτω τὰ περὶ ἕκαστον ἴδια θεωρεῖν) (189b31–33).

Now, we see Aristotle writing explicitly about the difference between what are commonly called in the literature, "accidental" and "substantial" change, which we noted just earlier that he had been foreshadowing in his differentiation between primary contraries and true contraries. Here he talks about change in terms of *gignomenon* (becoming) and not in terms of *kinêsis* or *metabole*. Still at the early stages of his inquiry, he is deriving the *archai* and has not yet introduced language of motion and change. Here, we see Aristotle first discussing becoming in the sense of predicative change, meaning both that there is something undergoing the change and that there is a result of the change.[24] This seems to be the type of becoming easily observed with the senses, e.g., the pale person becoming tanned or the man becoming properly trained.

When we talk about a man becoming properly trained, Aristotle asks whether we are talking really about the man becoming properly trained, about the untrained becoming properly trained, or about the untrained man becoming a properly trained man. He then wishes to point out that in the first two instances, i.e., the man becoming properly trained and the untrained becoming properly trained, what becomes is simple; whereas, in the third instance, i.e., when the untrained man

---

[24] *gignomenon* in this chapter is ambiguous—in some cases, it is the thing becoming and, in other cases it is that which results from the becoming. See Ross's translation of i 7 (1936, 345) and especially Wolfgang Wieland (1970, 113 fn 8 in "Introduction: The Study of Nature and the Nature of Time" and 14 in Chap. 3) who further explains the distinction: (1) ὃ γίγνεται, "terminus ad quem" or "das Resultat des Werdens" (that which results from the process of becoming) and (2) γιγνόμενον, "terminus a quo" or "dem Werdenden" (that which is becoming, that which will undergo the process) and explains that Aristotle's reader must decide for herself which of these Aristotle intends in any given instance. In wrestling with the ambiguity, Aristotle's intended meaning of "terminus a quo" is at stake. Aristotle's point here is that there is something becoming which is persisting through the becoming even though there is another something, which does not persist (189b32).

becomes properly trained, what becomes is complex (189b36–190a4).[25] There is a difference between the first two instances, however. Aristotle is quick to point out that while both are instances of simple things becoming, in the first instance, the simple thing, which is not a contrary, remains; whereas, in the second instance, the simple thing is destroyed. This is to say that the man himself remains in the qualitative change from untrained to properly trained. The substance remains with the acquisition of the accidental form. The quality of untrained, however, does not remain in the change to properly trained. For this reason, Aristotle concludes that there must always be an underlying third thing that is itself becoming. In addition to the contraries, there must be a thing that is not a contrary, which survives all becoming so that there is something that withstands alteration (ἐάν τις ἐπιβλέψῃ ὥσπερ λέγομεν, ὅτι δεῖ τι ἀεὶ ὑποκεῖσθαι τὸ γιγνόμενον) (190a13). Aristotle notes that this is one numerically but more than one with respect to its form (καὶ τοῦτο εἰ καὶ ἀριθμῷ ἐστιν ἕν, ἀλλ᾽ εἴδει γε οὐχ ἕν) (190a15), as there is both the substance and the accident. This third thing is the natural substantial being, i.e., the person becoming tanned after having been previously pale, or the man becoming properly trained from having been previously untrained.

But, talking about accidental or predicative change is not the only way to speak about change. This is where Aristotle is going to correlate his earlier discussion of different types of contraries with a discussion of different types of coming to be. Aristotle differentiates the way we commonly speak about becoming: sometimes we say, "come to be" (γίγνεσθαι) and sometimes we say, "come to be so-and-so" (τόδε τι γίγνεσθαι) (190a31–32). Our language correlates to the difference between substantial and accidental change, respectively. Substances "come to be" without predication, but they "come to be so-and-so" with regard to a change in quality, quantity, relation, time, or place (190a34–36 and cf *Categories* 4). This means that the becoming in the first instance is a generation; whereas, the becoming in the other cases is a motion from a privative to a positive form. Aristotle's point here is that with accidental change, a subject or substantial being is always presupposed. It is crucial to see where he is going with this and what he is beginning to build here. It may seem that this is an obvious point, but it has been overlooked by those who single out Aristotle's Treatise on Time and other later topical sections of the *Physics*. Without substantial being, accidental change, which has to do with alterations, quantities, place, relation, and time, does not exist. Likewise, it is accidental change that allows for a permanent component of not-being in substantial being.

Of course, as Aristotle adds, substances also come to be from something too. The implication here is that something never comes from nothing. But, this point,

---

[25]φαμὲν γὰρ γίγνεσθαι ἐξ ἄλλου ἄλλο καὶ ἐξ ἑτέρου ἕτερον ἢ τὰ ἁπλᾶ λέγοντες ἢ τὰ συγκείμενα. λέγω δὲ τοῦτο ὡδί. ἔστι γὰρ γίγνεσθαι ἄνθρωπον μουσικόν, ἔστι δὲ τὸ μὴ μουσικὸν γίγνεσθαι μουσικὸν ἢ τὸν μὴ μουσικὸν ἄνθρωπον ἄνθρωπον μουσικόν. ἁπλοῦν μὲν οὖν λέγω τὸ γιγνόμενον τὸν ἄνθρωπον καὶ τὸ μὴ μουσικόν, καὶ ὃ γίγνεται ἁπλοῦν, τὸ μουσικόν? συγκείμενον δὲ καὶ ὃ γίγνεται καὶ τὸ γιγνόμενον, ὅταν τὸν μὴ μουσικὸν ἄνθρωπον φῶμεν γίγνεσθαι μουσικὸν ἄνθρωπον.

which is a challenge to Parmenidean monism, does not undermine the distinction between substantial becoming and accidental change. Both types of becoming require something whence they emerge. Aristotle reminds us, for example, that both animals and plants come originally from seed (190b3).[26] There is a material whence the form comes to be. But, the real difference here is that the latter is conditional on the former having already occurred. *Kinêsis* (usually rendered "motion"), which is a type of *metabole* (usually, "change"), requires substantial natural being. In order for substance to exist, it must have come to be by way of generation, another kind of *metabole*. Aristotle makes this point when he explains that, "everything comes to be from both subject and form" (ὅτι γίγνεται πᾶν ἔκ τε τοῦ ὑποκειμένου καὶ τῆς μορφῆς) (190b19–20). The subject is the substance, which is composed of substantial matter and form, and the form is the accidental form. He reminds us of the properly trained man. The properly trained man is a complex thing; it is composed of a man, which is a subject, and the quality of proper training, the accidental form.

Thus, based on his own account, Aristotle is able to conclude that the *archai* are sometimes two and sometimes three (190b29–30). Formerly, it is the contraries themselves, the privative and the positive accidental form, which are the principles of nature. Later, the underlying subject of change is taken into account.

Having touched on the idea of privative form, Aristotle is ripe to discuss the place of non-being in the fundamental principles. He acknowledges the ancient quandary that it is impossible to understand the paradox of becoming: either a thing becomes from what is or from what is not. If something is not, nothing can come from it, and if something is, it already exists and can no longer come to be. Aristotle seems to return to a common sense argument based on experience—not confused perceptions, but analyzed and clear perceptions—that subjects can "be" in various ways. He uses the example of a doctor who exists already as a doctor, but he becomes other things apart from his identity as doctor (191b1–2). The doctor turns gray, not as a doctor, but as a dark haired thing. He builds a house, not as a doctor, but as a house builder. With these examples, Aristotle shows that the Pre-Socratic paradox of becoming is a problem not based on common sense or experience. Instead, it is based on a strict logic of non-contradiction, which ignores or is ignorant of the way natural beings actually exist. This is remedied by two important distinctions Aristotle makes with regard to the being of nature. On the one hand, there is this important difference between substance, or subject, and accidental form. I can exist as a woman and yet become many different things: wife, dog-owner, art collector, house-rehabber, etc. The predicates do not come from nothing; they come from their contraries, which Aristotle names a privative form. On the other hand, Aristotle notes a second explanation, which he has entertained in both *Metaphysics* iv 7 and viii, concerning the difference between being actually and

---

[26]ὅτι δὲ καὶ αἱ οὐσίαι καὶ ὅσα [ἄλλα] ἁπλῶς ὄντα ἐξ ὑποκειμένου τινὸς γίγνεται, ἐπισκοποῦντι γένοιτο ἂν φανερόν. ἀεὶ γὰρ ἔστι ὃ ὑπόκειται, ἐξ οὗ τὸ γιγνόμενον, οἷον τὰ φυτὰ καὶ τὰ ζῷα ἐκ σπέρματος.

Spontaneous generation is also possible (cf. HA vi 15, 569a13–19, 25–26, HA vi 16, 570a3–10, GA iii 11, 763a24–763b5), but it is not discussed here.

being potentially. At the heart of the being of nature is the potentiality to be in another way.

It is certainly not a coincidence that Aristotle spends so much time partially defending, but largely contending, with the *endoxa*. There was something basic that his predecessors got right: there is an interplay between contraries at the heart of nature's fundamental principles. But, what they did not get right was the absolute necessity of positing "the negative part" or privative form at the heart of all accidental change, and thus of the nature of natural beings. Aristotle is able to grasp what they did not because he has been willing in part to demonstrate his conclusions about the principles of the nature of natural beings based on experience. Aristotle explains what he takes to be Plato's oversight (192a4–15)[27]:

> Now we distinguish matter and privation, and hold that one of these, namely the matter, is not-being only in virtue of an attribute which it has, while the privation in its own nature is not-being; and that the matter is nearly, in a sense is, substance, while the privation in no sense is. They, on the other hand, identify their Great and Small alike with not being, and that whether they are taken together as one or separately. Their triad is therefore of quite a different kind from ours. For they got so far as to see that there must be some underlying nature, but they make it one-for even if one philosopher makes a dyad of it, which he calls Great and Small, the effect is the same, for he overlooked the other nature. For the one which persists is a joint cause, with the form, of what comes to be—a mother, as it were. But the negative part of the contrariety may often seem, if you concentrate your attention on it as an evil agent, not to exist at all.

Being and becoming are logically straightforward terms, but they are easily equivocated and/or oversimplified when not discussed in terms of what is possible and/or not possible for actual natural beings. In natural beings there are two senses of not-being, which necessarily exist as part of the being of natural beings. On the one hand, the underlying nature, or matter, of natural substantial beings, i.e., of the subject undergoing change, is always "not-being" in the sense that it is always "not-x, y, z" where x, y, z represent various qualitative predicates that it is only in potentiality but not in actuality. The man is not trained, but in being untrained, he is potentially trained. In this sense, not being is not actual non-existence in a substantial sense. Rather, not-being in this sense signifies a potentiality for that which is not-yet. The other sense of not-being is in the privation itself, which is half of the contrary and one of the principles or *archai* of the nature of natural beings. The privative form has no substantial existence. It is only when the privative form is understood in conjunction with the underlying subject, i.e., with the matter, that it

---

[27]ἡμεῖς μὲν γὰρ ὕλην καὶ στέρησιν ἕτερόν φαμεν εἶναι, καὶ τούτων τὸ μὲν οὐκ ὂν εἶναι κατὰ συμβεβηκός, τὴν ὕλην, τὴν δὲ στέρησιν καθ' αὐτήν, καὶ τὴν μὲν ἐγγὺς καὶ οὐσίαν πως, τὴν ὕλην, τὴν δὲ οὐδαμῶς οἱ δὲ τὸ μὴ ὂν τὸ μέγα καὶ τὸ μικρὸν ὁμοίως, ἢ τὸ συναμφότερον ἢ τὸ χωρὶς ἑκάτερον. ὥστε παντελῶς ἕτερος ὁ τρόπος οὗτος τῆς τριάδος κἀκεῖνος. μέχρι μὲν γὰρ δεῦρο προῆλθον, ὅτι δεῖ τινὰ ὑποκεῖσθαι φύσιν, ταύτην μέντοι μίαν ποιοῦσιν καὶ γὰρ εἴ τις δυάδα ποιεῖ, λέγων μέγα καὶ μικρὸν αὐτήν, οὐθὲν ἧττον ταὐτὸ ποιεῖ τὴν γὰρ ἑτέραν παρεῖδεν. ἡ μὲν γὰρ ὑπομένουσα συναιτία τῇ μορφῇ τῶν γιγνομένων ἐστίν, ὥσπερ μήτηρ ἡ δ' ἑτέρα μοῖρα τῆς ἐναντιώσεως πολλάκις ἂν φαντασθείη τῷ πρὸς τὸ κακοποιὸν αὐτῆς ἀτενίζοντι τὴν διάνοιαν οὐδ' εἶναι τὸ παράπαν.

becomes for the matter not-being with the potential for being. Aristotle's discussion of privative form here replaces the concept of potentiality, which is to say that the privation is at the same time potentiality; the untrained woman is at the same time potentially trained.

Aristotle further discusses the relationship between matter and privative form (192a25–33). Namely, he explains that the privative form is contained in the matter, and, as such, it both (1) comes to be and ceases not to be, i.e., in the sense that the privative form, when in the matter, is a potentiality for the positive form and this potentiality allows for the matter to constantly become what it is not-yet, and (2) it does not come to be and cease to be, i.e., substantially it remains what it is. Regarding the sense in which it does not come to be and cease to be, Aristotle explains that if it did substantially come to be and cease to be it would require a primary substratum, an underlying thing, itself. Since he defines matter as "the primary substratum of each thing, from which it comes to be without qualification, and which persists in the result" (λέγω γὰρ ὕλην τὸ πρῶτον ὑποκείμενον ἑκάστῳ, ἐξ οὗ γίγνεταί τι ἐνυπάρχοντος μὴ κατὰ συμβεβηκός) (192a32–33), it would not be possible that matter required a matter since the character of matter to provide this substratum is precisely what makes it special. Aristotle concludes this discussion asserting that an investigation into the first principle of form is outside of the scope of a science of nature (περὶ δὲ τῆς κατὰ τὸ εἶδος ἀρχῆς, πότερον μία ἢ πολλαὶ καὶ τίς ἢ τίνες εἰσίν, δι' ἀκριβείας τῆς πρώτης φιλοσοφίας ἔργον ἐστὶν διορίσαι, ὥστ' εἰς ἐκεῖνον τὸν καιρὸν ἀποκείσθω) (192a36–192b1). We might assume that this is precisely because the scientist has no empirical access to this subject, and natural science, as Aristotle has ably demonstrated to us, requires not just explanation, but demonstration. He then reaffirms the scope of his inquiry: the natural, i.e., perishable, forms (περὶ δὲ τῶν φυσικῶν καὶ φθαρτῶν εἰδῶν) (192b1). Put another way, Aristotle's primary subject here is natural beings in so far as they undergo accidental change.

In sum, Aristotle has set up three absolutely crucial points in *Physics* i: (1) becoming is a general term that we need to differentiate. Aristotle will do this primarily in *Physics* iii 1 and in *Physics* iv; (2) being is a general term that needs to be understood in terms of substantial and accidental form, or in terms of potentiality and actuality; (3) in order to understand these important distinctions in natural beings, we have to heed the external stimuli that puts our perceptual faculties into motion; otherwise, we could find ourselves making perfectly valid or sound arguments in the order of explanation that immediately do not follow when tested in the order of sense. Likewise, we need the order of explanation to flesh out our immediate and confused perceptions, lest we be unable to distinguish specific instantiations of a kind.

The final line of *Physics* i leads Aristotle's reader nearly to a re-beginning of the work. When Aristotle has finished his treatment of the *archai*, he announces that it is time for a "fresh start" (διωρίσθω ἡμῖν οὕτως πάλιν δ' ἄλλην ἀρχὴν ἀρξάμενοι λέγωμεν) (192b4). He begins *Physics* ii with his definition of nature, and he establishes for his reader what exactly sets apart a natural object—the subject of investigation—from a non-natural object. The difference between natural objects and non-natural objects is a difference in cause: natural objects come into being by way of "natural" causes, and non-natural objects come into being by way of other causes (τῶν ὄντων τὰ μέν ἐστι φύσει, τὰ δὲ δι' ἄλλας αἰτίας) (192b9). Natural objects have an inner principle of motion and rest with regard to the possibilities for accidental change: (1) with respect to place, (2) with respect to quality, (3) with respect to quantity (τούτων μὲν γὰρ ἕκαστον ἐν ἑαυτῷ ἀρχὴν ἔχει κινήσεως καὶ στάσεως, τὰ μὲν κατὰ τόπον, τὰ δὲ κατ' αὔξησιν καὶ φθίσιν) (192b13–22). Nature is something that inheres in something, and here Aristotle means this principle of motion and rest, i.e., of contraries and the substance to which these contraries are predicated. Nature, then, is the capability for self-locomotion, alteration, and increase/decrease, but this is only so because the being with these capabilities is a natural substantial being, a subject, which has come to be. Nature is two-fold. And, to investigate it, the physicist must attend both to the underlying subject and also to the way it changes (*kinêsis*).

1.3 The Role of *Kinêsis* in Nature

In *Physics* iii, Aristotle's reader begins to reap the benefit of understanding Aristotle's emphasis on both scope, access, and method and the potentiality/actuality distinction in the very being of natural beings in *Physics* i and how that led him to his definition of nature as an inner principle of motion and rest in *Physics* ii. The subject of motion, (κίνησις, *kinêsis*), in *Physics* iii comes directly and inextricably out of Aristotle's previous discussions. Aristotle writes: "Nature is a principle of motion and change, and it is the subject of our inquiry. We must therefore see that we understand what motion is; for if it were unknown, nature too would be unknown" (200b12–14).[28] If the scientist aims to know better the subject of her inquiry, i.e., the nature of natural beings, and if the nature of natural beings is an inner principle of motion and rest, then she ought to investigate what motion or accidental change (*kinêsis*) is and just how it happens.[29]

In addition to a discussion of *kinêsis*, Aristotle likewise intends to study those things that may be related to or conditions of *kinêsis*. Aristotle explains, "When we

---

[28]Ἐπεὶ δ' ἡ φύσις μέν ἐστιν ἀρχὴ κινήσεως καὶ μεταβολῆς, ἡ δὲ μέθοδος ἡμῖν περὶ φύσεώς ἐστι, δεῖ μὴ λανθάνειν τί ἐστι κίνησις· ἀναγκαῖον γὰρ ἀγνοουμένης αὐτῆς ἀγνοεῖσθαι καὶ τὴν φύσιν.

[29]See Aquinas's commentary (paragraph 276) on the importance of this opening argument for an understanding of motion's place in Aristotle's natural philosophy: "Nature is the principle of motion and change, as is evident from the definition set down in Book II. (But how motion and change differ, will be shown in Book V.) And thus it is evident that if one does not know motion, one does not know nature, since the former [motion] is placed in the definition of the latter [nature]. Since, therefore, we intend to present the science of nature, we must make motion understood." This is to say that knowledge of nature is impossible if we do not know motion. Or, put otherwise, we investigate motion because we aim to know about nature. I will emphasize this same argument as reason for the subsequent study of infinite, place, void, and time. We want to know them only because we want to know about nature.

have determined the nature of motion, our next task will be to attack in the same way the terms which are involved in it" (διορισαμένοις δὲ περὶ κινήσεως πειρατέον τὸν αὐτὸν ἐπελθεῖν τρόπον περὶ τῶν ἐφεξῆς) (200b15).[30] These terms involved with motion are thus involved with natural beings insofar as it is the nature of natural beings to move. He will deal with the infinite (ἄπειρον) because "motion is supposed to belong to the class of things which are continuous (δοκεῖ δ' ἡ κίνησις εἶναι τῶν συνεχῶν); and the infinite presents itself first in the continuous—that is how it comes about that 'infinite' is often used in definitions of the continuous; for what is infinitely divisible is continuous." After the infinite, Aristotle will deal with three additional terms, place, void, and time (τόπου καὶ κενοῦ καὶ χρόνου) because they "are thought to be necessary conditions of motion" (ἄνευ...κίνησιν ἀδύνατον εἶναι) (200b16–21).[31] It may be the case that Aristotle's claims about the relationship between motion and the infinite, place, void, and time, and thus his claim that he should treat these topics in his physics, are presuppositions from the *endoxa*. Ross (Ross 1936, 534) explains, for example, that, "It is not Aristotle's own opinion that motion implies a void; he does not believe in the existence of a void...the implication of a void is one of the *endoxa*, since it was insisted on by the atomists." Judging by the language alone, this conclusion is not clear. Aristotle uses the term, "δοκεῖ" to suggest a relationship between *kinêsis* and things that are continuous, but this could simply mean "it seems" based on first perceptions and is not necessarily a reference to what others believed. But, when we consider both (1) the likelihood that Aristotle's approach to investigating *kinêsis* will be parallel in method to his investigation of the *archai*, which did begin with an examination of the *endoxa*, and (2) some of Aristotle's predecessors did indeed espouse the relationship among motion and continuity, and motion, the infinite, place, void, and time, we might conclude that the impetus for Aristotle to treat the topics of the infinite, place, void, and time is indeed to contend with the *endoxa*. Accordingly, he will proceed to temper their explanations with demonstrations in order to come up with what appear to be the true conclusions.

Before moving on to speak first about *kinêsis*, Aristotle explains that the infinite, void, place, and time are all common to the present inquiry and asserts that they will each be dealt with in turn (δῆλον οὖν ὡς διά τε ταῦτα, καὶ διὰ τὸ πάντων εἶναι κοινὰ καὶ καθόλου ταῦτα, σκεπτέον προχειρισαμένοις περὶ ἑκάστου τούτων) (200b21–24). Hussey (1983, 56) thinks that Aristotle intends to include *kinêsis* here as those things common to the study of nature. For Hussey, Aristotle needed to justify the inclusion of these topics in his general work on nature. But, given that Aristotle has just asserted the certain relationship between nature and motion at 200b12–14, there seems no need to justify a discussion of *kinêsis*, in particular. We have just supposed

---

[30]The sense of this passage is that Aristotle will attempt to deal with the things that come after *kinêsis* insofar as they become topics for physics because *kinêsis* is a topic for physics. "Terms," as the ROT calls these things, is not a perfect way of talking about them. Nevertheless, for lack of a better name, I will refer to them as terms.

[31]A more literal translation here would be simply that without place, void, and time, motion is impossible (ἀδύνατον).

the other topics to be a carry over from the *endoxa*; Aristotle will later show us which of these are indeed appropriate for inclusion in his physics.

Aristotle introduces his discussion of *kinêsis* with a characterization of three different ways things exist: "(1) what exists in a state of fulfillment only (ἔστι δὴ [τι] τὸ μὲν ἐντελεχείᾳ μόνον), (2) what exists as potential, (3) what exists as potential and also in a state of fulfillment (τὸ δὲ δυνάμει καὶ ἐντελεχείᾳ), one being a 'this', another 'so much', a third 'such', and similarly in each of the other modes of the predication of being" (τὸ μὲν τόδε τι, τὸ δὲ τοσόνδε, τὸ δὲ τοιόνδε, καὶ τῶν ἄλλων τῶν τοῦ ὄντος κατηγοριῶν ὁμοίως) (200b25–28). We see at the start, then, how Aristotle's emphasis on the interplay between potentiality and actuality in the being and becoming of natural objects is going to have immediate application in his inquiry into the nature of *kinêsis*. Notice, though, that the second category above, "(2) what exists as potential" does not exist in the Greek.[32] Ross (1936, 534–535) explains "the absence of the first τὸ δὲ δυνάμει as due to haplography"; the scribe simply forgot to write the phrase again. If we can explain the exclusion of the phrase, and if Aristotle really did intend to distinguish a category of things that are just in potential, Ross notes that this is a departure from his general doctrine. This is to say, that what we see here is Aristotle deliberately adapting his general doctrine of potentiality, i.e., that the nature of potentiality is to be fulfilled, to his physics. Ross suggests that Aristotle does this in preparation to explain the infinite and the void. I will later argue that Aristotle's allowance here will help to explain his theory of time as well. Since Aristotle will shortly present us with his definition of *kinêsis*, Ross (1936, 535) believes it relevant that Aristotle here opposes the unchangeable and the changeable: "τὸ μὲν ἐντελεχείᾳ μόνον is that which is always actually what it ever is, in respect of substance, size, quality, and the other categories (b27–28); τὸ δὲ δυνάμει καὶ ἐντελεχείᾳ is that which passes from a state of potentiality to one of actuality in any of these respects." The difference between the two, then, is that the changeable is in part potentiality. Recall that Aristotle previously emphasized the aspect of non-being in matter, insofar as matter has the potential to change in any number of ways.

The differentiation Aristotle makes between the unchangeable and the changeable then brings him to discuss the nature of *kinêsis*. *Kinêsis* is not a substantial being itself with principles, elements, and causes to demarcate. Instead, he explains, "there is no such thing as motion over and above the things. It is always with respect to substance or to quantity or to quality or to place that what changes changes" (200b32).[33] *Kinêsis* is nothing over and above the natural beings; it describes the principle way of being for natural beings—as both *dunamis* and *entelecheia*. Motion is that which nature does, by its very definition. And, it is our observation of motion that alerts us to the nature of natural beings. The wording Aristotle holds onto in *Physics* i, of a more general notion of becoming

---

[32]According to the apparatus, τὸ δὲ δυνάμει does appear in the commentary tradition.

[33]οὐκ ἔστι δὲ κίνησις παρὰ τὰ πράγματα μεταβάλλει γὰρ ἀεὶ τὸ μεταβάλλον ἢ κατ' οὐσίαν ἢ κατὰ ποσὸν ἢ κατὰ ποιὸν ἢ κατὰ τόπον.

(*gignomenon*), exists over and above nature, but *kinêsis* in this context does not. When Aristotle famously defines motion in *Physics* iii first as "The fulfilment of what exists potentially, in so far as it exists potentially, is motion" (ἡ τοῦ δυνάμει ὄντος ἐντελέχεια, ᾗ τοιοῦτον, κίνησίς ἐστιν) (201a10–11), then as "It is the fulfillment of what is potential when it is already fully real and operates not as itself but as movable, that is motion" (ἡ δὲ τοῦ δυνάμει ὄντος <ἐντελέχεια>, ὅταν ἐντελεχείᾳ ὂν ἐνεργῇ οὐχ ᾗ αὐτὸ ἀλλ' ᾗ κινητόν, κίνησίς ἐστιν) (201a27–29), and finally as, "the fulfilment of the movable qua movable" (ἡ κίνησις ἐντελέχεια τοῦ κινητοῦ, ᾗ κινητόν) (202a7–8), he is not describing an abstract concept only tangentially related to nature.

How to understand Aristotle's use of the term *entelecheia* in the definition of *kinêsis* is debated in the literature, i.e., what Aristotle meant to convey in defining motion (in part) as *dunamis ontos entelecheia*. Deciding what *entelecheia* means here is crucial to recognizing the emphasis Aristotle is placing on potentiality in his definition of *kinêsis*. *Entelecheia* was traditionally translated, "actualization" (see for example Ross 1936, 537). This translation renders the first definition of *kinêsis*: "The actualization of what exists potentially, in so far as it exists potentially." For Kosman (1969, 40), "'Actualization' is an inelegant and in many ways misleading rendering of *entelecheia*" and understanding *entelecheia* in this way leads to, "two independent and unhappy accounts of Aristotle's definition of motion. On one account, Aristotle is understood to be defining motion as the actualization (process) of a potentiality into an actuality; on the other, he is understood to be defining motion as the actuality (product) of a potentiality to be in motion" (Kosman 1969, 45). According to Kosman (1969, 46), the way out of the traditional and problematic reading of *entelecheia* in Aristotle's definition of motion is to, "construct an account more svelte which (1) recognizes that Aristotle's definition talks about the actuality of a potentiality, (2) recognizes that potentiality as a potentiality to be, e.g., the potentiality of bricks and stones to be a house, but (3) yields motion and not its result, i.e., the act of building and not the house which is its product." Sachs (2010, 8), who agrees with Kosman, also illustrates the importance of highlighting the potential aspect of *kinêsis*: "The growth of the puppy is not the actualization of its potentiality to be a dog, but the actuality of that potentiality as a potentiality." The emphasis here is on potentiality as a kind of being, an actual state of being. Kosman (1969, 56) wants us to understand Aristotle's definition of motion therefore as: "the functioning, the full manifesting of a potentiality qua potentiality, or more precisely, the functioning of a being which is potential as that potential being." Aristotle's definition of *kinêsis* calls attention to the realness of potentiality—when something is in motion, it is *actually* a potentiality and not only a process of becoming something else.[34]

In his discussion of the definition of *kinêsis* (201a10–202b29) Aristotle elaborates on the relationship between potentiality/actuality in natural beings, emphasizing the character of potentiality in the ways these beings exist, and explains the

---

[34]Broadie (1982), Hussey (1993), and Coope (2008) all defend this view as well.

relationship between the mover and the moveable in *kinêsis*. He demonstrates these concepts with three paradigmatic examples, (1) the subject of sickness and health, (2) the house being built, and (3) the simultaneity of teaching and learning. He begins then with the potentiality/actuality relationship: "the same thing can be both potential and fulfilled, not indeed at the same time or not in the same respect, but e.g., potentially hot and actually cold" (201a20–21).[35] Highlighting the aspect of potentiality, so crucial to *kinêsis*, Aristotle solves the Pre-Socratic problem of non-contradiction. The fire can be hot and cold at the same time, for example, if we understand it to be potentially one and actually the other. Aristotle is harking back to his argument in *Physics* i 5 where he defended the Pre-Socratic claim that principles are contraries. Here, he advances this defense, but he now shows explicitly (previously, he referenced *Meta. theta*) that contraries can exist simultaneously, given that they are understood to exist in two different potencies. When the woman is actually monolingual, she is potentially bilingual. But, the implication here is that she is not at rest being monolingual; rather, she is in motion (or, potentially in motion) on her way to being bilingual. She is actually potentially bilingual.

It is this reasoning that allows Aristotle to connect his arguments about *kinêsis* to his argument from *Physics* i 7 about the fundamental principles of nature. If the *archai* of natural objects are three, and if nature is an inner principle of motion and rest, then he needs to show that the principles somehow require *kinêsis*. After all, *kinêsis* is so integrally important to nature that it is only by way of *kinêsis* that one knows the nature of natural beings and yet only by way of things that *kinêsis* exists. When he is explaining the definition of *kinêsis*, then, it is no surprise that Aristotle shows contraries to exemplify perfectly the way *kinêsis* works with the principle of non-contradiction. He demonstrates this point with the first of three paradigmatic examples in this section:

> To be capable of health and to be capable of illness are not the same; for if they were there would be no difference between being ill and being well. Yet the subject both of health and of sickness—whether it is humour or blood—is one and the same (201a33–201b2).[36]

Again, we see the three principles, which were argued for in *Physics* i: two contraries and the underlying subject. Whatever is said to be healthy or ill stays the same, but the qualities of actual health and potential illness are two, i.e., they are not the same. To be actually one and potentially the same are two separate qualities. If they were two parts of the same quality, then Aristotle could have been content calling the contrary principles, "one" instead of "two."

---

[35]ἐπεὶ δ' ἔνια ταὐτὰ καὶ δυνάμει καὶ ἐντελεχείᾳ ἐστίν, οὐχ ἅμα δὲ ἢ οὐ κατὰ τὸ αὐτό, ἀλλ' οἷον θερμὸν μὲν ἐντελεχείᾳ ψυχρὸν δὲ δυνάμει.

[36]δῆλον δ' ἐπὶ τῶν ἐναντίων τὸ μὲν γὰρ δύνασθαι ὑγιαίνειν καὶ δύνασθαι κάμνειν ἕτερον-καὶ γὰρ ἂν τὸ κάμνειν καὶ τὸ ὑγιαίνειν ταὐτὸν ἦν—τὸ δὲ ὑποκείμενον καὶ τὸ ὑγιαῖνον καὶ τὸ νοσοῦν, εἴθ' ὑγρότης εἴθ' αἷμα, ταὐτὸν καὶ ἕν.

The potentiality exists as an actuality itself, not as an actualization. Aristotle demonstrates this in terms of his famous house-building example (201b6–15):

> For each thing of this kind is capable of being at one time actual, at another not. Take for instance the buildable as buildable. The actuality of the buildable as buildable is the process of building. For the actuality of the buildable must be either this or the house. But when there is a house, the buildable is no longer buildable. On the other hand, it is the buildable which is being built. The process then of being built must be the kind of actuality required. But building is a kind of motion, and the same account will apply to the other kinds also.[37]

The actuality of the process of building exists simultaneously with, yet is different from, the potentiality for the materials to become a house. In the process of building, the house does not fully exist in actuality; it is as yet incomplete. The house exists in potentiality even in the building materials, and the process of building with these materials is the actuality of this potentiality.

In the background here is another distinction between the relations that make motion possible, namely the relationship between the mover and the moveable. Without both, there can be no *kinêsis*. We know that natural objects are capable of *kinêsis*, for motion means nothing over and above these objects. For this reason, Aristotle writes that "motion is in the moveable. It is the fulfillment of this potentiality by the action of that which has the power of causing motion" (202a12–15). The potentiality for *kinêsis*, is always already in that which can be moved. When there is motion, there is an actuality of this potentiality, which aligns with the actuality of that which moves the moveable. The actuality of the potentiality for what can be moved and the actuality of that which moves are thus simultaneous but in the sense that they are two sides of one coin, i.e., they are both directed toward one end. Aristotle indicates this unusual relationship between mover and moved (202a15–20), which is to say that they share a single actuality. He famously claims that "the steep ascent and the steep descent are one."[38]

Again, Aristotle puts himself in a situation where he seems to be advocating a logical impossibility, contrary to what the *endoxa* would have argued. He goes on to explain, however, that while there is a sense in which the action of the mover (agent) and the moveable (patient) are one and the same, they are fundamentally different. The difference lies in the source, i.e., the potentiality and actuality for moving something lies in the agent, and the potentiality and actuality of movement lies in the patient. And the source relates directly to the sense in which Aristotle means actuality here. Namely, actuality of two things can be the same in any given instant, if actuality is meant in two different senses. This is the same argument we saw Aristotle advancing in *Physics* i 7 with regard to contraries. He is playing with

---

[37]γὰρ ἕκαστον ὁτὲ μὲν ἐνεργεῖν ὁτὲ δὲ μή, οἷον τὸ οἰκοδομητόν, καὶ ἡ τοῦ οἰκοδομητοῦ ἐνέργεια, ἢ οἰκοδομητόν, οἰκοδόμησίς ἐστιν (ἢ γὰρ οἰκοδόμησις ἡ ἐνέργεια [τοῦ οἰκοδομητοῦ] ἢ ἡ οἰκία ἀλλ' ὅταν οἰκία ᾖ, οὐκέτ' οἰκοδομητὸν ἔστιν οἰκοδομεῖται δὲ τὸ οἰκοδομητόν ἀνάγκη οὖν οἰκοδόμησιν τὴν ἐνέργειαν εἶναι) ἡ δ' οἰκοδόμησις κίνησίς τις. ἀλλὰ μὴν ὁ αὐτὸς ἐφαρμόσει λόγος καὶ ἐπὶ τῶν ἄλλων κινήσεων.

[38]καὶ τὸ ἄναντες καὶ τὸ κάταντες ταῦτα γὰρ ἓν μέν ἐστιν, ὁ μέντοι λόγος οὐχ εἰς ὁμοίως.

the principle of non-contradiction in that he is showing that there is a fundamental difference in being between potential and actuality.

Aristotle appeals to the example of teaching and learning (202b3–10). Teaching requires a relationship between someone teaching and someone being taught. The teacher has a potentiality to teach and the learner has the potentiality to be taught. In this scenario, the teacher is the agent, and the learner is the patient. When the teacher actually teaches the student, the effect is that the student actually learns something. To say that teaching and learning is the same does not mean that to learn and to teach are the same thing; rather, the teaching and the learning are occurring simultaneously. The learning is happening because the teaching is happening, but the teaching could not be happening if the potential for learning were not already in the learner, nor if the learner were absent. Learning happens in so far as the agent/ patient relation between teacher and student exists in actuality.

Aristotle rounds out his discussion of *kinêsis* with a sample definition of qualitative motion: "alteration is the fulfillment (*entelecheia*) of the alterable as alterable (or, more scientifically, the fulfillment of what can act and what can be acted on, as such)" (202b24–26). He then concludes that all kinds (e.g., quantitative, locomotive) of motion will be defined in a similar fashion.

1.4 From *Kinêsis* to *Chrónos*

In *Physics* iii 4 Aristotle turns to the first common term thought to be associated with *kinêsis*, the infinite (ἄπειρον, *apeiron*). He will then go on to discuss place, void, and time. As he transitions from his discussion of *kinêsis* to his investigation of these topics, he echoes his statements at 200b21–24 that these subjects are of concern to the science of nature. He confirms that his support for this comes from the tradition; all previous natural philosophers considered the infinite to be not only a topic of their study, but also a principle of beings (202b30–203a4).[39]

We must consider carefully Aristotle's study of the infinite, as it will be with all subsequent topics thought to be related to motion, within the context of his previous arguments. Aristotle has essentially argued that natural science must take account of the principle role potentiality plays in the being of natural self-subsistent beings. Whereas, this fact may have eluded an a priori philosopher of nature, Aristotle has used examples from empirical experience to show that potentiality, while not always actualized, is actual (*entelecheia*) in nature. If we do not understand the interplay between potentiality and actuality, we are at a loss for understanding the *archai* of nature, the distinction between Aristotle's three types of beings—actuality only, potentiality only, and that which is both potentiality and actuality (cf. 200b25–28),

---

[39]σημεῖον δ᾽ ὅτι ταύτης τῆς ἐπιστήμης οἰκεία ἡ θεωρία ἡ περὶ αὐτοῦ· πάντες γὰρ οἱ δοκοῦντες ἀξιολόγως ἧφθαι τῆς τοιαύτης φιλοσοφίας πεποίηνται λόγον περὶ τοῦ ἀπείρου, καὶ πάντες ὡς ἀρχήν τινα τιθέασι τῶν ὄντων.

the definition of *kinêsis*, and so too, the important distinction between moveable and moved. If we miss the relative novelty of what he has been arguing regarding potentiality and actuality, especially that he sets himself apart from the *endoxa* in the emphasis he places on potentiality—both that there are beings that are just potentiality and that the potentiality inherent in the matter of natural beings allows them to be what they are, i.e., to change, then we fall into the trap of the Pre-Socratic physicists who saw only contradiction in the face of topics where something could exist alongside its contrary. What Aristotle is going to show here is that these terms thought to be associated with, and perhaps conditions (cf. 200b20) of *kinêsis*, thought to be so crucial in the study of nature, exist largely, if not always, as is the case with infinity and void, as potentiality. He is going to steer us away from thinking of the infinite, place, void, and time as being in some sense self-subsistent natural beings (οὐσίαν αὐτὸ ὄν) themselves.

Aristotle will treat the infinite first precisely because it was previously thought to be an *archai* of beings, and if motion, place, void, and time are all beings, then we might believe them to be infinite beings. Aristotle will oppose the view that the infinite is an *archai* of beings and argue that the infinite is in fact nothing "actual" at all. Likely due to its relative importance for understanding the other terms thought to be associated with *kinêsis*, Aristotle's treatise on the infinite is the longest when compared with his treatises on place, void, and time.

His treatment of the infinite begins with a line of questioning that we will find standard in his entire treatment of the terms of *kinêsis*: whether or not there is such a thing as the infinite (202b35–36).[40] This question of course recalls the doctrine of the *endoxa*, which has taught that motion, infinity, void, place, and time are all actual self-subsistent natural beings qua themselves. Aristotle asks whether there is such a being, for each in turn, as a rhetorical question because he will show that there is not such a being in the sense that the *endoxa* has agreed that there is. He will follow up that question inquiring into the manner of its existence and what it is.[41]

Aristotle's treatise on the infinite, not unlike his previous arguments for (1) the *archai* of the nature of natural beings and (2) the definition of *kinêsis*, will first consider the explanation of the *endoxa* and then add demonstration to arrive at true conclusions about the science of nature.

---

[40]Heinemann (2012, 5) rightly suggests that Aristotle's approach here follows *Posterior Analytics* ii 1, 89b24–5: "We seek four things: the fact, the reason why, if it is, what it is" (ζητοῦμεν δὲ τέτταρα, τὸ ὅτι, τὸ διότι, εἰ ἔστι, τί ἐστιν.). Heinemann writes, "Aristotle's point in asking the question as to "if it is" is just to secure some subject matter of inquiry to exist.".

[41]At the start of his discussion of *kinêsis*, we do not find this question. The question as to whether motion exists was not asked, since it had been previously established that the principle of nature was an inner principle of motion and rest (192b13–22). "Nature is a principle of motion and change, and it is the subject of our inquiry. We must therefore see that we understand what motion is; for if it were unknown, nature too would be unknown" (Ἐπεὶ δ' ἡ φύσις μέν ἐστιν ἀρχὴ κινήσεως καὶ μετα βολῆς, ἡ δὲ μέθοδος ἡμῖν περὶ φύσεώς ἐστι, δεῖ μὴ λανθάνειν τί ἐστι κίνησις ἀναγκαῖον γὰρ ἀγνοουμένης αὐτῆς ἀγνοεῖσθαι καὶ τὴν φύσιν) (200b12–14). Thus, it was clear that motion exists. Aristotle needed then to establish in what way it existed.

The Pythagoreans and Plato both held the infinite to be a "self-subsistent substance" (οὐσίαν αὐτὸ ὄν) instead of an attribute of something else (συμβεβηκός τινι) (203a4–5). The Pythagoreans believed that the infinite could be found in natural objects accessible to the scientist's investigation; the Platonists agreed with this general idea and for the latter it could be found both in these objects as well as in the Forms. For the Pre-Socratic physicists, the infinite is likened to the divine, reported by Aristotle to have been thought of as 'immortal and imperishable' (ἀθάνατον καὶ ἀνώλεθρον) (203b13–14). Aristotle dismisses these ideas and cites five plausible arguments for the existence of the infinite: (1) from time; (2) from the division of magnitudes; (3) as the source of all generation and corruption; (4) against ultimate limits, since limits are always relative; (5) with regard to number, mathematical magnitudes, and that which is outside the heavens—infinite body (203b15–25). He goes on to show problems present themselves whether one is arguing for or against the existence of the infinite. First of all, there are different senses in which the infinite might be said to exist, e.g., as a substance, as an accident, or as something else altogether. He will begin by arguing against that the infinite can exist as substance separate from natural objects (204a8–29).[42] On the one hand, if the infinite is a substance, but neither a magnitude (μέγεθος), nor an aggregate (πλῆθος), it will be indivisible (ἀδιαίρετον); magnitudes and aggregates are of course always divisible. But, what is indivisible is not infinite in the sense meant by the *endoxa*, i.e., the sense that Aristotle means it at this point, as 'that which cannot be gone through.' So, it seems impossible that the infinite is a substance; there is a contradiction between the type of substance the infinite *qua* infinite *can* be logically and the way it is discussed. As a further problem, the infinite is known to be an attribute of both number (ἀριθμὸς) and magnitude; but, neither number nor magnitude are substances. Aristotle reasons that the infinite cannot be a substance when it describes that which is not itself substance.[43] He then returns to the idea that if the infinite were a substance that it would be indivisible. His additional argument here is that if it were divisible, it would be divided into

---

[42]Χωριστὸν μὲν οὖν εἶναι τὸ ἄπειρον τῶν αἰσθητῶν, αὐτό τι ὂν ἄπειρον, οὐχ οἷόν τε. εἰ γὰρ μήτε μέγεθός ἐστιν μήτε πλῆθος, ἀλλ' οὐσία αὐτό ἐστι τὸ ἄπειρον καὶ μὴ συμβεβη κός, ἀδιαίρετον ἔσται (τὸ γὰρ διαιρετὸν ἢ μέγεθος ἔσται ἢ πλῆθος) εἰ δὲ τοιοῦτον, οὐκ ἄπειρον, εἰ μὴ ὡς ἡ φωνὴ ἀόρατος. ἀλλ' οὐχ οὕτως οὔτε φασὶν εἶναι οἱ φάσκοντες εἶναι τὸ ἄπειρον οὔτε ἡμεῖς ζητοῦμεν, ἀλλ' ὡς ἀδιεξίτητον. εἰ δὲ κατὰ συμβεβηκὸς ἔστιν τὸ ἄπειρον, οὐκ ἂν εἴη στοιχεῖον τῶν ὄντων, ᾗ ἄπειρον, ὥσπερ οὐδὲ τὸ ἀόρατον τῆς διαλέκτου, καίτοι ἡ φωνή ἐστιν ἀόρατος. ἔτι πῶς ἐνδέχεται εἶναί τι αὐτὸ ἄπειρον, εἴπερ μὴ καὶ ἀριθμὸν καὶ μέγεθος, ὧν ἐστι καθ' αὐτὸ πάθος τι τὸ ἄπειρον; ἔτι γὰρ ἧττον ἀνάγκη ἢ τὸν ἀριθμὸν ἢ τὸ μέγεθος. φανερὸν δὲ καὶ ὅτι οὐκ ἐνδέχεται εἶναι τὸ ἄπειρον ὡς ἐνεργείᾳ ὂν καὶ ὡς οὐσίαν καὶ ἀρχήν ἔσται γὰρ ὁτιοῦν αὐτοῦ ἄπειρον τὸ λαμβανόμενον, εἰ μεριστόν (τὸ γὰρ ἀπείρῳ εἶναι καὶ ἄπειρον τὸ αὐτό, εἴπερ οὐσία τὸ ἄπειρον καὶ μὴ καθ' ὑποκειμένου), ὥστ' ἢ ἀδιαίρετον ἢ εἰς ἄπειρα διαιρετὸν πολλὰ δ' ἄπειρα εἶναι τὸ αὐτὸ ἀδύνατον (ἀλλὰ μὴν ὥσπερ ἀέρος ἀὴρ μέρος, οὕτω καὶ ἄπειρον ἀπείρου, εἴ γε οὐσία ἐστὶ καὶ ἀρχή) ἀμέριστον ἄρα καὶ ἀδιαίρετον. ἀλλ' ἀδύνατον τὸ ἐντελεχείᾳ ὂν ἄπειρον ποσὸν γάρ τι εἶναι ἀναγκαῖον.

[43]Note here that Aristotle states without argument that number is not a substance. This is one of the clear indications that he does not think about number as a self-subsistent being qua itself. This is a clue as to how to understand Aristotle's definition of time, which classifies time as a number.

parts that would be themselves infinite, and there cannot be multiple infinities. He concludes again that the infinite must be indivisible, but then concedes that if the infinite were to exist in full completion qua substance, it would have to be a definite quantity, i.e., divisible. The upshot of Aristotle's arguments here is that the infinite cannot be an actual thing, substance, or principle on account of the various paradoxes, which result from such an idea.

He next considers that the infinite could exist accidentally, as a predicable being, quickly showing that this is impossible, as that which is accidental cannot be an *archai* (204a30–31).[44] A principle, as he argued earlier (203b4–5), is a source itself, which by definition cannot be traced back further to a more primordial source. Thus, Aristotle successfully contends with the *endoxa*—in this case the Pythagoreans whose characterization of the infinite Aristotle shows to be self-contradictory: "With the same breath they treat the infinite as substance, and divide it into parts" (ἅμα γὰρ οὐσίαν ποιοῦσι τὸ ἄπειρον καὶ μερίζουσι).

Finally, Aristotle wonders whether the infinite could exist in an alternative way, e.g., "present in mathematical objects and things which are intelligible and do not have extension" (204a35–204b1).[45] There is suddenly a bit of confusion, however, as Aristotle wants to stick to the aim of his treatise and subject of his inquiry, i.e., to natural objects. Again, he works to show that there are no natural objects known by sense that can increase infinitely. A body cannot be infinite, if we define body as "bounded by a surface" (204b5–6) and infinite as "boundlessly extended" (204b20), nor can number since for Aristotle and the Greeks (see Klein 1969), number is not a symbolic expression, but that which is numerable.[46] A second argument concludes that while the infinite body must be either compound or simple—as all bodies must—it cannot be either (204b11–24).[47] Aristotle thus seems to rule out completely that the infinite is any kind of body at all.

---

[44]ἀλλ' εἰ οὕτως, εἴρηται ὅτι οὐκ ἐνδέχεται αὐτὸ λέγειν ἀρχήν, ἀλλ' ᾧ συμβέβηκε, τὸν ἀέρα ἢ τὸ ἄρτιον.

[45]εἰ ἐνδέχεται ἄπειρον καὶ ἐν τοῖς μαθηματικοῖς εἶναι καὶ ἐν τοῖς νοητοῖς καὶ μηδὲν ἔχουσι μέγεθος.

[46]As Ross (1936, 541) notes, "When Aristotle says (*Met.* 987b27) that the Pythagoreans identified real things with numbers, it is not to be supposed that they reduced reality to an abstraction, but rather that they did not recognize the abstract nature of numbers. What they were doing was little more than to state that the characteristics of things depended, to a large extent, on the number and the numerical relations of their components." Hussey (1983, 88) reminds us that Aristotle is only talking about positive integers here.

[47]οὔτε γὰρ σύνθετον οἷόν τε εἶναι οὔτε ἁπλοῦν. σύνθετον μὲν οὖν οὐκ ἔσται τὸ ἄπειρον σῶμα, εἰ πεπερασμένα τῷ πλήθει τὰ στοιχεῖα. ἀνάγκη γὰρ πλείω εἶναι, καὶ ἰσάζειν ἀεὶ τἀναντία, καὶ μὴ εἶναι ἐν αὐτῶν ἄπειρον (εἰ γὰρ ὁποσῳοῦν λείπεται ἡ ἐν ἑνὶ σώματι δύναμις θατέρου, οἷον εἰ τὸ πῦρ πεπέρανται, ὁ δ' ἀὴρ ἄπειρος, ἔστιν δὲ τὸ ἴσον πῦρ τοῦ ἴσου ἀέρος τῇ δυνάμει ὁποσαπλασιονοῦν, μόνον δὲ ἀριθμόν τινα ἔχον, ὅμως φανερὸν ὅτι τὸ ἄπειρον ὑπερβαλεῖ καὶ φθερεῖ τὸ πεπερασμένον) ἕκαστον δ' ἄπειρον εἶναι ἀδύνατον σῶμα μὲν γάρ ἐστι τὸ πάντη ἔχον διάστασιν, ἄπειρον δὲ τὸ ἀπεράντως διεστηκός, ὥστε τὸ ἄπειρον σῶμα πανταχῇ ἔσται διεστηκὸς εἰς ἄπειρον. ἀλλὰ μὴν οὐδὲ ἓν καὶ ἁπλοῦν εἶναι σῶμα ἄπειρον ἐνδέχεται, οὔτε ὡς λέγουσί τινες τὸ παρὰ τὰ στοιχεῖα, ἐξ οὗ ταῦτα γεννῶσιν, οὔθ' ἁπλῶς.

Nonetheless, he provides additional arguments as to why the infinite cannot be a sensible body. A sensible body has many things predicated of it: quantity, quality, place, relation. What is infinite is not predicated in these ways, as categories are limits and the infinite is by definition unlimited. Finally, he concludes: "It is plain from these arguments that there is no body which is actually infinite" (206a7–8).[48] Thus, he begins again to consider that the infinite exists in some other way. It must be the case that it does, Aristotle claims, as the consequences of its non-existence are impossible to imagine (Ὅτι δ' εἰ μὴ ἔστιν ἄπειρον ἁπλῶς, πολλὰ ἀδύνατα συμβαίνει): (1) that there is a beginning and end to time; (2) a magnitude will not be divisible into magnitudes; (3) number will not be infinite (206a9–11).[49] What this amounts to, then, is that while, per Aristotle's arguments, there is a sense in which the infinite does not exist, i.e., as a sensible body, substance, or predicate, there is also a sense in which the infinite does exist. When we consider this in terms of the division of being Aristotle initially set out—actuality only, potentiality only, and that which is both potentiality and actuality (see 200b25–28)—we can rule out two possibilities. The infinite is in no way actuality, neither actuality itself, nor actuality and potentiality. Therefore, the infinite must exist as potentiality exclusively.[50]

The difference between potentiality in the case of the infinite, as opposed to the ways in which any substantial being can potentially change itself to be predicated by any contrary from whatever is a current actual predication (e.g., an untrained man is potentially trained), is that whatever is infinite will never become fulfilled. Aristotle illustrates this point when he admonishes, "but we must not construe potential existence in the way we do when we say that it is possible for this to be a statue—this will *be* a statue, but something infinite will not be in actuality" (206a19–21).[51] The infinite is only ever actual in potentiality; there is no point of actualization. Rather, what Aristotle means by "infinite" is similar to the sense of being one intends when saying, "it is day or it is the games" (206a22). Aristotle is thus demonstrating his claim that the infinite is not a substantial being, principle, or sensible body; it is not a subject, i.e., "a this" (τὸ ἄπειρον οὐ δεῖ λαμβάνειν ὡς τόδε τι), but "a process of coming to be or passing away" (ἀλλ' ἀεὶ ἐν γενέσει ἢ φθορᾷ, πεπερασμένον) (206a29–33).[52] Aristotle thus describes the infinite as that which, "has this mode of existence: one thing is always being taken after another, and each

---

[48]ὅτι μὲν οὖν ἐνεργείᾳ οὐκ ἔστι σῶμα ἄπειρον, φανερὸν ἐκ τούτων.

[49]τοῦ τε γὰρ χρόνου ἔσται τις ἀρχὴ καὶ τελευτή, καὶ τὰ μεγέθη οὐ διαιρετὰ εἰς μεγέθη, καὶ ἀριθμὸς οὐκ ἔσται ἄπειρος. ὅταν δὲ διωρισμένων οὕτως μηδετέρως φαίνηται ἐνδέχεσθαι, διαιτητοῦ δεῖ, καὶ δῆλον ὅτι πὼς μὲν ἔστιν πὼς δ' οὔ.

[50]It is for this reason that one may say that for Aristotle there is a sense in which the infinite does not exist; see for example Heinemann (2012, 5).

[51]Οὐ δεῖ δὲ τὸ δυνάμει ὂν λαμβάνειν, ὥσπερ εἰ δυνατὸν τοῦτ' ἀνδριάντα εἶναι, ὡς καὶ ἔσται τοῦτ' ἀνδριάς, οὕτω καὶ ἄπειρον ὃ ἔσται ἐνεργείᾳ.

[52]Following Ross, 206a29–34 is bracketed in the ROT. Since Ross excises the bracketed sentence as an alternative version of 206a18–29 (ROT, 351), I refer to the sentence seemingly out of order to help explain what Aristotle is saying at 206a23–25.

thing that is taken is always finite, but always different" (206a27–29).[53] In this sense, what Aristotle describes as infinite names an intrinsic aspect of the nature of natural beings—there is ever the possibility of accidental change as long as they exist.

To the extent the infinite is for Aristotle a characterization of a potentiality of a thing, the infinite, like *kinêsis*,[54] never exists over and above natural subsistent being. Sorabji (1983, 210) is right to point out here that Aristotle's account of the infinite is highly original, as it defines the infinite in terms of the finite. According to Sorabji, there are two upshots to Aristotle's argument: (1) "infinite is connected with a *process*" (his emphasis) and, (2) "infinity is always what has something outside of it." This is to say that the infinite is a potential aspect of the nature of natural being, and as such, always exists in conjunction with these things. This, as Sorabji notes, is a view antithetical to those of Aristotle's predecessors who thought that the infinite was, "something which is so all-embracing that it has nothing outside of it." The ways in which something can be said to be always, "taken after another…always different" are clearly numerous. Unsurprisingly, then, we can talk about different sorts of things as being "infinite," and we will find that different sorts of things are infinite in likewise different ways (cf. 207b22). For example, *kinêsis* is called infinite because, in the case of locomotion, the ground covered is always finite but each time different. Similarly with alteration and growth, what changes is finite, but there is always more change to come until the substance ceases to be. Time is said to be infinite in the sense that it is of *kinêsis* (207b24).

In addition, Aristotle talks about the infinite in terms of "the potentially infinite," in the sense that beings can be potentially endlessly divided (206b5–6, 206b17–18). If infinite division is possible, in what sense could this be so? Sorabji (1983, 211) gives two possible interpretations: (1) Aristotle thinks that these infinite divisions are actually materially possible, and (2) Aristotle's infinity means the recognition of endless potential division, what Sorabji names "the finitist view." I am in agreement with Sorabji, who assigns the latter to Aristotle, arguing that, "certainly, Aristotle would allow only a finite number of *actually* existing divisions." But, Sorabji questions whether or not this would be the case for "*potentially* existing divisions" (his emphasis). He allows that the question is ambiguous, and, after weighing possible replies, concludes that, "Aristotle cannot there [in the *Physics*] afford to admit any collections which are more than finite, if his analysis of infinity is to surmount the problems which it is intended to surmount."

Aristotle's account of the infinite is an account of possibility; it controversially argues that the infinite *is* only ever potentiality. It was covered first because of the Pre-Socratic assumption that it was an *archē* of other being. Now that Aristotle has disabused his reader of such a notion, he will move on to a discussion of the other terms. Let his treatment of the infinite be a paradigm of sorts for what is to come;

---

[53]Ὅλως μὲν γὰρ οὕτως ἔστιν τὸ ἄπειρον, τῷ ἀεὶ ἄλλο καὶ ἄλλο λαμβάνεσθαι, καὶ τὸ λαμβανόμενον μὲν ἀεὶ εἶναι πεπερασμένον, ἀλλ' ἀεί γε ἕτερον καὶ ἕτερον.

[54]Recall 200b33: "There is no such thing as motion over and above the things.".

Aristotle is defeating the views of the *endoxa* with an eye to pointing out the role of potentiality in nature.

Aristotle begins *Physics* iv with an inquiry into the being of place (τόπος, *topos*). The first question is expected: whether there is such a thing as place (208a28). And, this is our indication that Aristotle will now deal with the *endoxa*.

Aristotle tells us that people suppose everything is somewhere and nothing is nowhere. Certainly, as he reminds us, locomotion requires place. In order to change places, place must exist in some way. So too, it seems that place exists over and above the things in place because according to the *endoxa*: (1) the contents of a place can be re-placed by different content, e.g., water in a vessel can be emptied and replaced by air (208b6–7); (2) elements seem to have a proper place, as fire rises and earth descends (208b9–10); (3) the theory of void requires a theory of place since void is place without body (208b25). It is interesting to note as well that in Aristotle's discussion of proper place, he makes a distinction between relative place, or position, and proper place, or power. He says that to us, "up and down," "left and right" change given our relative position; whereas, they remain the same in nature. "Downward," for example, is a force or power evidenced by the fact that some things by nature move toward the earth, i.e., down. This is not an episte-mological/metaphysical distinction, however, as Aristotle is only pointing out that our perspective is sometimes at odds with nature's perspective. This echoes his claim from *Physics* i 1, that some things are clearer to nature, while others are clearer to us. Part of studying nature, when one is a part of nature, is to recognize this difference without inappropriately singling oneself out as somehow outside of the bounds of the investigation.

Other concerns about the possibility that place is a body itself include: (1) if it were, there would be two bodies in the same place (209a5–7); (2) place is not the cause of anything else (209a20–22); (3) if everything has a place, then place would have a place and so on ad infinitum (209a23–25); (4) place has a size, but its size must grow since there are things in place, which themselves grow (209a26–28). So, Aristotle sets out first to say whether place might be matter or form. If it were the former, it would be the extension itself of the magnitude, and if it were the latter, it would be the limit of the body in place (209b1–10). He then cites Plato as the only thinker to actually try and say *what* place is and not just *that* it is. According to Aristotle's reading, Plato, in the *Timaeus,* claims that matter is identical to space because the space that a body is in is the body itself.[55] Aristotle concludes saying that place is neither matter, nor form. Matter and form, as we have seen, are inseparable from a natural object. Place, on the other hand, is separable from the

---

[55] As Hussey (1983, 105) remarks, this is a careless reading of Plato's *Timaeus* 48e–52d. Aristotle seems to have left out sufficient differences between his idea of matter (*hyle*) and Plato's receptacle (*chóra*), thus embellishing the similarities needed for a proper analogy. Hussey explains, "Aristotle interprets Plato's receptacle as playing the same role as Aristotelian matter...the existence and whereabouts of a piece of Aristotelian matter are always dependent on those of the body of which it is the matter, whereas the Platonic receptacle seems to be an independent entity of which the parts cannot change their relative positions.".

object (209b21–30). This then leads Aristotle to the new premise that place is a vessel or container. The implication of such a view is that place is indeed something, and this points Aristotle's inquiry to ask *what sort* of thing it is. Important to note here is that Aristotle is explicit about the fact that place is something that, although separable from natural objects themselves, is not a natural object itself.[56] This will begin with a look to the meaning of "in" with regard to what it might mean to be "in" something, e.g., "in place." Aristotle highlights a function of "in," that as Hardie and Gaye point out, does not quite capture the Greek preposition, 'ἐν,' which is in use here. The sense here is the way we mean "in" when we say, "in a vessel"; it usually means "inside" (Πάντων δὲ κυριώτατον τὸ ὡς ἐν ἀγγείῳ καὶ ὅλως ἐν τόπῳ) (210a24). Aristotle brings in examples from experience, which marks a brief turn from what has been mostly arguments from the "order of explanation," in contention with the *endoxa* to show that a thing containing other things does not have to be either the form, the matter, or the same thing as that which is contained (210b7–30). In *Physics* iv 4, he will go on to explain the sense in which place is a vessel.

Aristotle begins by stating his assumptions: (1) that place is what contains that of which it is the place, and is no part of the thing; (2) that the primary place of a thing is neither less nor greater than the thing; (3) that place can be left behind by the thing and is separable; (4) that all place admits of the distinction of up and down; (5) each of the bodies carried to its appropriate place and rests there, and this makes the place either up or down (210b36–211a5). And, then, he returns to the previous point that the topic of place has come up only because there is motion with respect to place (211a11–12). Since locomotion is a movement from one place to another, place becomes a topic for the natural scientist. It is not a topic *qua* itself; it is requisite for locomotion. Since the heavens are in constant movement themselves, Aristotle concludes that they must also be in place. Another type of *kinêsis* relevant to place is change in quantity, as increase and diminution require change in size of place.

Aristotle raises a puzzle in which he creates an analogy between place and the underlying thing, or *hypokeimenon*, which he discussed in *Physics* i. Place, he argues seems to remain as natural objects change place. A vessel has air at point α, which is replaced by water at point β. The place inside the vessel remains. The analogy is only partially effective, however, as underlying matter neither separates from the natural object, nor does it contain the object (211b30–212a2). Place, thus, is "the boundary of the containing body at which it is in contact with the contained body" (ἀνάγκη τὸν τόπον εἶναι τὸ λοιπὸν τῶν τεττάρων, τὸ πέρας τοῦ περιέχοντος σώματος <καθ' ὃ συνάπτει τῷ περιεχομένῳ>) (212a6–7), or, "the innermost motionless boundary of what contains it" (ὥστε τὸ τοῦ περιέχοντος πέρας ἀκίνητον πρῶτον, τοῦτ' ἔστιν ὁ τόπος) (212a20). And, finally, "If then a body has another body outside it and containing it, it is in place, and if not, not" (ᾧ ι μὲν οὖν σώματι ἔστι τι ἐκτὸς σῶμα περιέχον αὐτό, τοῦτο ἔστιν ἐν τόπῳ, ᾧ δὲ μή, οὔ) (212a31–32).

---

[56]Instead, recall that it is in some sense an attribute of motion and relation (200a3–4).

Aristotle then retakes up a topic he had suggested earlier; namely, that some things are potentially in place, while other things are actually in place. Things are potentially in place when they are parts of a homogenous continuous substance (212b5–6). They are actually in place when they are separated but in contact (212b6–7). And, some things are per se in place, while others are accidentally in place. The former includes all bodies that move from one place to another or bodies that increase or decrease in size. The latter includes the soul, which since it is contained in a body, is only ever in place by virtue of the fact that the body is in place.

Aristotle concludes his treatise on place asserting once again that only moveable bodies are in place. This challenges the view of the *endoxa* that place is a substantial being itself. Place exists, but not *qua* itself. Instead, place serves as the limit of a moveable body, only potentially existent unless a body exists to help actualize it. Place can be anywhere, so long as a moveable body is there, too. If something is not movable, either intrinsically, i.e., natural objects, or by propulsion, e.g., artifacts, then it is not "in place" (212b27–30).[57]

Aristotle's final move in his discussion of place is to suggest that it is an analogue to matter. First, he starts with a place-whole analogy: "that which is in place has the same relation to its place as a separable part to a whole, as when one moves a part of water or air; so, too, air is related to water, for the one is like matter, the other form—water is the matter of air, air as it were the actuality of water" (212b36–213a21).[58] We can imagine a container filled with water. The space inside the container is the whole, and the water is a separable part of it. When the space is filled with water, the container and the contained appear to be one. There is no part yet to be filled. But, when some of the water escapes, there is a vacant part, which is filled with air. As parts of the whole, the water is potentially air. When any more water escapes, the vacant place will come to contain air; the water will be re-placed by the air. As both Hardie and Gaye and Aquinas, in his commentary (Aquinas 1961, 239), instruct, Aristotle will not fully explain the relationship between the elemental bodies until his more narrow work of natural philosophy on generation and corruption. Here, he concludes, "for water is potentially air, while air is potentially water, though in another way…if the matter and the fulfillment are the same thing (for water is both, the one potentially, the other in fulfillment), water will be related to air in a way as part to whole. That is why these have contact: it is organic union when both become actually one" (213a3 and 213a5–10).[59] According

---

[57]καὶ ἔστιν ὁ τόπος καὶ πού, οὐχ ὡς ἐν τόπῳ δέ, ἀλλ' ὡς τὸ πέρας ἐν τῷ πεπερασμένῳ. Οὐ γὰρ πᾶν τὸ ὂν ἐν τόπῳ, ἀλλὰ τὸ κινητὸν σῶμα.

[58]καὶ γὰρ τὸ μέρος, τὸ δὲ ἐν [τῷ] τόπῳ ὡς διαιρετὸν μέρος πρὸς ὅλον ἐστίν, οἷον ὅταν ὕδατος κινήσῃ τις μόριον ἢ ἀέρος. Οὕτω δὲ καὶ ἀὴρ ἔχει πρὸς ὕδωρ.

[59]οἷον ὕλη γάρ, τὸ δὲ εἶδος, τὸ μὲν ὕδωρ ὕλη ἀέρος, ὁ δ' ἀὴρ οἷον ἐνέργειά τις ἐκείνου· τὸ γὰρ ὕδωρ δυνάμει ἀήρ ἐστιν, ὁ δ' ἀὴρ δυνάμει ὕδωρ ἄλλον τρόπον…ἀσαφῶς δὲ νῦν ῥηθὲν τότ' ἔσται σαφέστερον. εἰ οὖν τὸ αὐτὸ [ἡ] ὕλη καὶ ἐντελέχεια (ὕδωρ γὰρ ἄμφω, ἀλλὰ τὸ μὲν δυνάμει τὸ δ' ἐντελεχείᾳ), ἔχοι ἂν ὡς μόριόν πως πρὸς ὅλον. διὸ καὶ τούτοις ἁφὴ ἔστιν· σύμφυσις δέ, ὅταν ἄμφω ἐνεργείᾳ ἓν γένωνται.

to *De Generatione et Corruptione*, water is potency to air simply. The air is likewise potentially water, because water could be added to re-place the air. Water is the matter, ready at any time to take on the form of air. It is potentially air, and yet it is water. As water, it has become one with its place in the vessel, i.e., with air.

Aristotle's treatise on void (κενόν, *kenon*) follows his analytic on place quite naturally. After all, void is thought to be a place without a body (213b31). Aristotle shows that if body is presupposed to be something tangible, with properties of heavy and light, then by deduction it appears that there are places with nothing in them (214a1–5). But, as he has just shown, the elemental bodies are "in place" because they move and change. If place is a limit of a body, and if bodies that are immoveable are not in place, and if all moveable bodies are in place, then there can be no place without a body (216a24–26). The concept of void, in fact, can only exist in a conception where place is believed to be a thing separate from the bodies it contains. If place is a self-subsistent natural being itself, then there could be places that exist without containing anything. These places would be void, i.e., empty. Similarly, if we are founding our conclusions on experience in the world, we can imagine observing any given space occupied by many objects. We might say that the areas in the space where objects exist are filled places; whereas, the places where no objects exist could be considered "empty places." This is of course to ignore the air that is in place around the substantial objects. So, too, it may suggest again that one is thinking of place as something separate from body, which can be filled or unfilled. Place for Aristotle is a container; but, it is a container in which the contained and the container become one. It is the limit of the thing "in place." Thus, there is no separate void (ὅτι μὲν τοίνυν οὐκ ἔστι κεχωρισμένον κενόν, ἐκ τούτων ἐστὶ δῆλον) (216b20).

After considering the *endoxa* that has void existing as the source of movement,[60] (217b20–22), Aristotle concludes that the only sense in which void could be said to exist is "the matter of the heavy and the light, qua matter of them" (οὕτω δ' ἡ τοῦ βαρέος καὶ κούφου ὕλη, ᾗ τοιαύτη, εἴη ἂν τὸ κενόν) (217b23), meaning that "void" is the name for the very tension existing naturally among the *archai* of nature. Void would be then the possibility that some substantial being could change accidental form as a matter of nature, i.e., a being that, like the infinite, is a being only in potentiality.[61] If we do not want to name this aspect of nature, "void," then void exists neither actually, nor potentially in Aristotle's physics.

That concludes our examination of Aristotle's scope, access, goals, and method in the *Physics*, his arguments for the number and kind of principles in nature, and his definitions of nature, motion, the infinite, place, and void in *Physics* i–iv 9. I

---

[60](cf ἐκ δὴ τῶν εἰρημένων φανερὸν ὡς οὔτ' ἀποκεκριμένον κενὸν ἔστιν, οὔθ' ἁπλῶς οὔτ' ἐν τῷ μανῷ, οὔτε δυνάμει, εἰ μή τις βούλεται πάντως καλεῖν κενὸν τὸ αἴτιον τοῦ φέρεσθαι).

[61]Coope (2008, 57 n22) does not seem to recognize the sense in which void can exist potentially. She asserts without argument that void does not exist for Aristotle and refers her reader to the treatise on void.

have stressed that Aristotle's emphasis on the interplay between potentiality and actuality in nature, and especially that he allows the modality of potentiality a unique ontological status to which he ends up assigning the infinite, and, in a sense, void, significantly sets him apart from the *endoxa*. In the next chapter, we will complete the transition from *kinêsis* to *chrónos* with an examination of Aristotle's (1) puzzles of time, and (2) his analytic of time, to include his definition of time.

Chapter 2
Physics iv 10-11 as a Parallel Account

Given the context of the *Physics* just explored, it will not be a surprise if time (*chrónos*) in Aristotle's analytic of time turns out to be not a being *qua* itself but an attribute of motion,[1] an interval.[2] First, let us take seriously 218a1, a rather neglected line in the treatise,[3] where Aristotle qualifies two types of time—*ho apeiros chrónos*, or infinite time, and *ho aei lambanamenos chrónos*, or taken time—suggesting that *chrónos* is homonymous—naming two different senses of time. It should be highly unlikely that *chrónos* in the *Physics* means infinite time—in brief, infinite time is outside the scope of the *Physics*, in so far as the *Physics* is principally concerned with the nature of natural beings and the allusions to infinite time seem relegated to *Physics* iv 10, where Aristotle works through the doxa. Thus, I turn to Chap. 11, what I call Aristotle's "analytic of time," where Aristotle first defines *chrónos*, to

---

[1]Interpreting *Physics* iv 11 is difficult, and the literature is divided on interpretation. I agree with Shoemaker 1969, Sorabji 1983, Hussey 1981 that time for Aristotle requires perception of *kinêsis*. Roark (2011, 56) claims that readers of Aristotle in this camp have not defended why Aristotle would hold this view here in the Treatise on Time but nowhere else. My defense is twofold: (1) I read the Treatise on Time as highly contextualized and parallel in structure to Aristotle's foregoing arguments about the terms of *kinêsis*. Time, like the infinite, place, and void is not considered a being qua itself in Aristotle's philosophy of nature here in the *Physics*. In short, this is an account of time relevant to an inquiry into the being of natural beings, i.e., an account of time taken. Recall, the Treatise on Time may have been the end of Aristotle's initial work on nature; (2) It is not the case that Aristotle does not at least assume this view in other works of his natural philosophy. I will look to some of these works in the final chapter.

[2]Already, in the very idea of "the time taken," there is a nod to the fact that time requires a "taker." Otherwise, time cannot be "taken." This seems a foreshadowing of the subsequent arguments about time and the soul in *Physics* iv 14.

[3]Namely, while there is literature discussing the difference between the two Greek times, *chrónos* and *kairos* (see for example Moutsopoulos 2010; Smith 1969), thus an acknowledgment that there was more than one sense of temporality for the Greeks, there has been no sustained discussion about the fact that *chrónos* itself seems to be a homonym—naming two different senses of time.

© The Author(s) 2015
C.C. Harry, *Chronos in Aristotle's Physics*,
SpringerBriefs in Philosophy, DOI 10.1007/978-3-319-17834-9_2

argue that time for Aristotle is a time interval, insofar at its actual existence depends on the motion of natural beings; it is not an a priori or self-subsistent being.[4]

Chapter 10 of the Treatise on Time is analogous in purpose to the initial chapters of each foregoing treatise, e.g., on the *archai* of nature, *kinêsis*, the infinite, place, and void.[5] Namely, it serves to discuss the *endoxa* as preparatory to Aristotle's actual analytic of time, which begins at 219a1–3: "It is evident, then, that time is neither movement nor independent of movement. We must take this as our starting-point and try to discover—since we wish to know what time is—what exactly it has to do with movement" ('Ότι μὲν οὖν οὔτε κίνησις οὔτ' ἄνευ κινήσεως ὁ χρόνος ἐστί, φανερόν).[6] In this chapter, I trace the development of Aristotle's analytic from this starting point until he both defines *chrónos* at 219b1 (ἀριθμὸς κινήσεως κατὰ τὸ πρότερον καὶ ὕστερον) and then, after some argument, reaffirms his definition at 220a25 ('Ότι μὲν τοίνυν ὁ χρόνος ἀριθμός ἐστιν κινήσεως κατὰ τὸ πρότερον καὶ ὕστερον). I attempt to show, by way of a proposal that the "now" for Aristotle is not only (1) non-temporal, as Coope (2005, 29) has suggested, but also (2) a referent for existing self-subsistent natural beings undergoing *kinêsis*, i.e., a referent to their modality, that the best reading of this analytic is to understand Aristotle's position on time to be that time is only ever potentially actual, and by consequence only ever potentially a continuum, unless it is apprehended as such. I support this reading in part by contrasting the way Aristotle dismisses that time could be a self-subsistent being composed of actual parts in *Physics* iv 10, and then argues that time is in some sense continuous, i.e., presumably a whole composed of parts, in Chap. 11.[7] I treat *Physics* iv 10 in both Sects. 2.1, and 2.2, in Sect. 2.3, I consider *Physics* iv 11.

---

[4]Ultimately, I agree with Roark that the before and after is non-temporal in Aristotle's account, thus with Coope that the business of numbering the before and after entails counting "nows," implying that "now" too is non-temporal; but, I will depart from Coope insofar as she argues that, "... there must be some *other* continuum, prior to time, on which the now depends for its existence" and that the other continuum is change, and instead propose that the other continuum— in line with the greater context of Aristotle's *Physics*—is a "this," the self-subsistent existing natural beings, "the matter" undergoing the change.

[5]Coope (2005, 17) also mentions the similarity in structure between the beginning of Aristotle's Treatise on Time and the way he began his account of place (209a2) and his account of the infinite (iii 4–5), but adds in n. 1 that while puzzles about the infinite are answered by Aristotle (iii 8), he wrongly claims that he has solved all of the puzzles about place at 212b22–23. Coope refers her reader to Ross (1936, 564).

[6]Roark (2011, 53) supports the theory that *Physics* iv 11 begins Aristotle's analytic of time, in Roark's words, "Aristotle's positive account of time."

[7]I offer a reading of *Physics* 11 despite that the order of arguments is challenging to understand in a coherent way (see for example Hussey (1983, 145) on the strange arrangement of the section).

2.1 Introducing the Issue of Time

Aristotle begins his Treatise on Time as he did with the other terms of motion see Chap. 1 fn 26; he will examine the *endoxa* and attempt to understand the difficulties of his subject—here, time (217b29–30).[8] Commentators commonly refer to such difficulties as the "paradoxes" or "puzzles" (*aporiai*) of time:[9]

(1) Does time exist or not?
(2) What is the nature of time?

Aristotle first considers the arguments for the non-existence of time. Or, if not the non-existence of time, the relative obscurity of whatever time is (ὅτι μὲν οὖν ἢ ὅλως οὐκ ἔστιν ἢ μόλις καὶ ἀμυδρῶς, ἐκ τῶνδέ τις ἂν ὑποπτεύσειεν) (218a1–2). He implies that time is a whole composed of parts when he brings up the commonly known "parts" (μέρη) of time: past and future. Past does not exist because it "has been and is not," and the other part "is going to be and is not" (τὸ μὲν γὰρ αὐτοῦ γέγονε καὶ οὐκ ἔστιν, τὸ δὲ μέλλει καὶ οὔπω ἔστιν) (218a2–3). But, then, curiously, Aristotle backtracks to state that any time "is made up of these" (218a4). Aristotle continues to argue that since in order for something divisible to exist it is necessary that all or some of its parts exist, but then seemingly exempts time from this conditional saying: "but of time some parts have been, while others are going to be, and no part of it is, though it is divisible" (τοῦ δὲ χρόνου τὰ μὲν γέγονε τὰ δὲ μέλλει, ἔστι δ' οὐδέν, ὄντος μεριστοῦ) (218a5–6). For Plato, in the *Timaeus*, days, nights, months, and years are all parts (μέρη) of time; the past (what "was") and future (what "will be") are not parts, but forms (εἴδη) of time (37e). It is thus unclear, if Aristotle is appealing to *endoxa* here, the source of the idea that "past" and "future" are parts of time. If Aristotle is not appealing to *endoxa*, the argument is circular. This is to say that if Aristotle is positing non-existent parts of time as a premise whence to conclude that time does not exist, he has already assumed that time is a whole, thus is composed of parts. The idea that time is a whole is problematic when we consider the arguments Aristotle has just made with regard to the kind of being he attributes to the infinite, place, and void. These are terms of *kinêsis* and not actual self-subsistent beings. Why then might Aristotle begin his Treatise on Time with the assumption that time is a whole?[10]

---

[8]Ἐχόμενον δὲ τῶν εἰρημένων ἐστὶν ἐπελθεῖν περὶ χρόνου· πρῶτον δὲ καλῶς ἔχει διαπορῆσαι περὶ αὐτοῦ καὶ διὰ τῶν ἐξωτερικῶν λόγων, πότερον τῶν ὄντων ἐστὶν ἢ τῶν μὴ ὄντων, εἶτα τίς ἡ φύσις αὐτοῦ.

[9]Coope (2005, 17) adds Aristotle's subsequent question, "What is time's relation to the present, or 'now'?" to the puzzles.

[10]Aristotle will argue in *Physics* 11 that time is continuous. Since the essence of continuity for Aristotle is that something is a whole with parts, that these parts are touching, and that there is the potential for infinite divisibility of the whole, it makes sense that he begins with this assumption. But, if his Treatise on Time is an investigation in the same vein as his previous queries into the terms of *kinêsis,* i.e., in the form of APo ii 1, 89b24–5, and beginning with *endoxa* in the order of explanation and proceeding to demonstrate that the term of motion is not a self-subsistent being

Aristotle clarifies that there are two ways to think about time: (1) infinite time (ἄπειρος χρόνος), and (2) time taken (λαμβανόμενος χρόνος) (218a1). Now, Aristotle has already shown that the infinite exists only to the extent that the potentiality for it exists, e.g., in the possibility for infinite divisibility. What are we then to make of the idea of "infinite time," mentioned here without explanation or definition?[11] Aristotle's reference to infinite time could mean two things: (1) a reference to the 'time" of his predecessors, that is, to Platonic time, whose emphasis on number may be traced back to the Pythagoreans,[12] or (2) an idea, whether from Plato or elsewhere, presupposed about the possibility for endless time (*aion*),[13] given that certain heavenly motions seem to be ceaseless and that the possibility exists (at least in the intellective faculty of the soul) for motion qua motion to continue forever.[14]

(Footnote 10 continued)

itself, this assumption seems impetuous. If Aristotle's puzzles are not just rhetorical, how can we assume something is continuous when we have not yet established whether or not it exists? Indeed, in her reply to Miller (1974, 139–141), Coope (2005, 20) raises a similar point when she says that Miller's suggestion that the puzzles of time could have been solved if Aristotle had said, "to be is to be surrounded by time" would not work because assuming that being is surrounded by time is to already assume that time exists, and whether or not time exists is the question Aristotle poses. Yet, Coope does not raise this same issue with regard to Aristotle's assumption that time is a whole composed of parts, i.e., is continuous.

[11]Coope (2005, 81) cites *Generation and Corruption* (338b9–11) to argue that Aristotle elsewhere posits "a pretemporal order that is both infinite and (in the relevant sense) linear," and she believes that Aristotle could have used this notion in the *Physics* to provide a temporal basis for the before and after, thus defending "his assumption about time's linearity." I will discuss shortly that Aristotle did not need a temporal basis for the before and after in his account of time in the *Physics* and that in fact before and after are not inherently temporal concepts.

[12]In the *Timaeus* 37d–38c, Plato defines time (*chrónos*) as a type of number: as the number according to which the universe, or Living Creature, moves (ποιεῖ μένοντος αἰῶνος ἐν ἑνὶ κατ' ἀριθμὸν ἰοῦσαν αἰώνιον εἰκόνα, τοῦτον ὃν δὴ χρόνον ὠνομάκαμεν) (37d) and as that which "imitates eternity and circles according to number" (κατ' ἀριθμὸν κυκλουμένου γέγονεν εἴδη) (38a). Later, he affirms that there are numbers of time (38c). So, he appears to be inconsistent regarding the relationship between time and number. The universe, or "Living Creature" has a mostly eternal nature, but cannot be fully eternal due to the fact that it was created. That which comes into being must also perish from being. So, it is said to have been modeled after eternity; yet, it is truly sempiternal. As such, despite having been generated, it will be for all time. As Helena Keizer (Keizer 1999, 88) points out, Aristotle seems to be referring to the *Timaeus* 38c1–3 in *De Caelo* i 10, 280a28–32. Here, Aristotle questions the idea that something can be both generated and existing for all time. In short, Aristotle calls into question the whole notion of creation. Cf. also *Physics* viii 1 251b15–20 where Aristotle challenges Plato's claim that time was created.

[13]Keizer (1999, 90) highlights the sense in which *aion* cannot be endless, i.e., it is "a completeness which is an end (*telos*) in all its fullness."

[14]Plato makes the connection between motion and time already in the *Timaeus* when he concludes that these things becoming in the world of sense do so in time. Time (*chrónos*) is the circling number, which imitates eternity (*aion*) (ἀλλὰ χρόνου ταῦτα αἰῶνα μιμουμένου καὶ κατ' ἀριθμὸν κυκλουμένου γέγονεν εἴδη) (38a).

If time as infinite refers to that which is unchanging and not becoming, it is not the kind of time we would expect Aristotle to discuss in the *Physics*.[15] We have seen his emphasis on becoming from the beginning of the work. Contrast that with the fact that there has been no mention whatsoever about the unchanging movement of the heavens.[16] Indeed, it would be beyond the access permitted to the natural scientist. This ever-continuous time is not the time, which is a term of *kinêsis* insofar as it refers to the nature of natural beings. Instead, it might be a subject for a more speculative thinker, perhaps a cosmologist. In both cases, then, the idea for infinite time (ἄπειρος χρόνος) is outside the scope of Aristotle's arguments here in the *Physics*; they would be beyond the scope, access, method, and goals of this inquiry.[17] Instead, Aristotle will focus on time that becomes an issue for us because it is a term of *kinêsis*—the time of this sort is a time interval—time taken (λαμβανόμενος χρόνος).

2.2 Eschewing the *Endoxa*

First, Aristotle investigates what appears to be a third part of time, the "now." But, "now," what we commonly think of as the present tense of time, is not going to be a part of time for Aristotle. Parts, he instructs, are measures of wholes, and parts themselves have parts (218a6–7). But, time for Aristotle is not made of nows, at least in the temporal sense. Aristotle is rejecting the idea that time could be represented as a string of points. We could imagine a string of beads to illustrate this commonly held view of time. Placing a finger on one bead isolates it as the "present"—whatever beads exist to the left of the finger are "the past," and the beads to the right are "the future." In one's actual experience of life, the now seems elusive. When can it be said actually to occur? Is it now? Now? Now? How about, now? No, it is always already gone. The future slides into the past before we can really acknowledge it. It takes great intention to experience each moment as it arrives.

But, this is not at all how Aristotle is thinking of "now," precisely because for him time is not going to end up being a linear continuum existing as a subsistent being itself, independent of natural objects. The treatment of what are commonly

---

[15]Though some have argued that *aion* is timeless (cf. Sorabji 1983, 126 n. 122 where he mentions von Leyden 1964; Keizer 1999, 89), Sorabji (1983, 126–127) appeals to *De caelo* i 9, 279a12–b3 to argue that Aristotle does not mean "timelessness" when he writes *aion*; but, rather, "everlasting duration." This is not to say, as Sorabji concludes, that Aristotle considers "possessers of this sort of *aion*" to be in time. Instead, Sorabji notes the "special sense" of time that Aristotle presents in the *Physics*.

[16]Aristotle will of course famously broach this topic in *Physics* viii, but one could argue that, in the spirit of many of Aristotle's treatises, the topics of the last book are preparatory to a subsequent topic of study. On this reading, Aristotle prepares us for the *de Caelo* at the end of the *Physics*.

[17]Sorabji (1983, 126) has noted that it does not seem that infinity can be a number. When this conclusion is then accepted as a premise here, since time is going to end up being a number (*arithmos*) for Aristotle, the idea that time could be both a number *and* infinite is self-contradictory.

held to be "parts of time," i.e., past, present, and future, then is meant to show the absurdity of understanding time in this way—if not the absolute illogicality, at least that such an understanding of time does not derive from the preceding theory of nature. Aristotle easily demonstrates that the past and future do not *actually* exist, i.e., we can clearly think about them, but they cannot be perceived, and now Aristotle sets out to understand "now."

He writes that the now seems to be bound by past and future and then wonders whether it is always the same or each time different (218a9–10). The arguments he then puts forth to show that neither is possible are not arguments made in earnest. On the contrary, he is disclosing the logical inconsistencies required to understand the present, "the now," as an actual part of time *qua* self-subsistent being. After giving arguments against each possibility, he concludes that there are "difficulties about the attributes of time" (περὶ μὲν οὖν τῶν ὑπαρχόντων αὐτῷ τοσαῦτ' ἔστω διηπορημένα) (218a30).[18] At this point in the text, this conclusion is easy to infer. There are clearly internal inconsistencies with the position that holds time to be a whole, composed of two parts that do not exist, and the now, which is not a part but acts as a marker between the two parts that do not exist, and is neither always the same, nor always different. Aristotle is peeling us away from holding the traditional view of time as a being itself, presupposed in our common understanding of nature and nudging us toward an internally consistent, sound, view of time as the "time taken." The reasoning here is the same type of reasoning Aristotle employs to explain accidental change. Things neither stay the same nor are ever different. Because in the *endoxa*, the now appears to be a part of time, or it is commonly talked about as if it were, Aristotle has to debunk this notion. Before demonstrating by analogy the impossibility of the now as ever same or as ever different, he defends the view in terms of the *endoxa*, i.e., as if the now were a being *qua* itself, a part of the whole of time (218a12–29).[19] Since Aristotle has established that the "parts" of time are not simultaneous—the past always has been and the future always will be, the now has always just ceased to be. We can verify this with experience. *When is* the now? Is it now? Now? Now? So, if the now has always just

---

[18]"Attributes" is not a perfect translation of τῶν ὑπαρχόντων, literally "posessions."

[19]ὁ δὲ χρόνος οὐ δοκεῖ συγκεῖσθαι ἐκ τῶν νῦν. ἔτι δὲ τὸ νῦν, ὃ φαίνεται διορίζειν τὸ παρελθὸν καὶ τὸ μέλλον, πότερον ἓν καὶ ταὐτὸν ἀεὶ διαμένει ἢ ἄλλο καὶ ἄλλο, οὐ ῥάδιον ἰδεῖν. εἰ μὲν γὰρ αἰεὶ ἕτερον καὶ ἕτερον, μηδὲν δ' ἐστὶ τῶν ἐν τῷ χρόνῳ ἄλλο καὶ ἄλλο μέρος ἅμα (ὃ μὴ περιέχει, τὸ δὲ περιέχεται, ὥσπερ ὁ ἐλάττων χρόνος ὑπὸ τοῦ πλείονος), τὸ δὲ νῦν μὴ ὂν πρότερον δὲ ὂν ἀνάγκη ἐφθάρθαι ποτέ, καὶ τὰ νῦν ἅμα μὲν ἀλλήλοις οὐκ ἔσται, ἐφθάρθαι δὲ ἀνάγκη ἀεὶ τὸ πρότερον. ἐν αὐτῷ μὲν οὖν ἐφθάρθαι οὐχ οἷόν τε διὰ τὸ εἶναι τότε, ἐν ἄλλῳ δὲ νῦν ἐφθάρθαι τὸ πρότερον νῦν οὐκ ἐνδέχεται. ἔστω γὰρ ἀδύνατον ἐχόμενα εἶναι ἀλλήλων τὰ νῦν, ὥσπερ στιγμὴν στιγμῆς. εἴπερ οὖν ἐν τῷ ἐφεξῆς οὐκ ἔφθαρται ἀλλ' ἐν ἄλλῳ, ἐν τοῖς μεταξὺ [τοῖς] νῦν ἀπείροις οὖσιν ἅμα ἂν εἴη· τοῦτο δὲ ἀδύνατον. ἀλλὰ μὴν οὐδ' αἰεὶ τὸ αὐτὸ διαμένειν δυνατόν· οὐδενὸς γὰρ διαιρετοῦ πεπερασμένου ἓν πέρας ἔστιν, οὔτε ἂν ἐφ' ἓν ᾖ συνεχὲς οὔτε ἂν ἐπὶ πλείω· τὸ δὲ νῦν πέρας ἐστίν, καὶ χρόνον ἔστι λαβεῖν πεπερασμένον. ἔτι εἰ τὸ ἅμα εἶναι κατὰ χρόνον καὶ μήτε πρότερον μήτε ὕστερον τὸ ἐν τῷ αὐτῷ εἶναι καὶ ἑνὶ [τῷ] νῦν ἐστιν, εἰ τά τε πρότερον καὶ τὰ ὕστερον ἐν τῷ νῦν τῳδί ἐστιν, ἅμα ἂν εἴη τὰ ἔτος γενόμενα μυριοστὸν τοῖς γενομένοις τήμερον, καὶ οὔτε πρότερον οὔτε ὕστερον οὐδὲν ἄλλο ἄλλου.

ceased to be, then that means it did exist. But, in what sense could it have existed? If it did not cease to be in itself, it might have ceased to be in another now. But, then, the now would be simultaneous with another now, which is impossible if both the now is a part of time and the parts of time do not exist. If "nows" cannot be simultaneous, it follows that when the present now "is," the prior now must have ceased to be.

But, neither can the now be always the same. Aristotle argues that since no determinate divisible thing has a single termination despite the ways it is extended, and, since the now is like a point, indivisible, the now is a termination. So too, it is possible to cut off a determinate time. And, as a negative account, if the now were always the same, what happened in the past would be simultaneous, i.e., "now," with what has happened subsequently. Therefore, the now cannot be always the same.

The arguments against the possibility for ever-different nows suggest, on the one hand, that the "now" does not actually exist. From the argument, since the now never actually existed, the prior now cannot have ceased to be in itself. The now cannot cease to be in itself because this entails that it must have existed. But, Aristotle never denies that the now exists. As we noted, it seems demonstrable by way of perception, even if the perception requires intention, to show that it does. But, he understands it as akin to a point, i.e., without parts itself. If it cannot have parts itself, then it cannot be part of a whole. Thus, as we saw, it is not a part of time.

Aristotle thus intends that the "now" is neither always the same, nor always different. Following which, he openly dismisses the "difficulties" dealt with in this preliminary chapter and remarks that, "the traditional accounts give us as little light as the preliminary problems which we have just worked through" (ὁμοίως ἔκ τε τῶν παραδεδομένων ἄδηλόν ἐστιν, καὶ περὶ ὧν τυγχάνομεν διεληλυθότες πρότερον) (218a31–32).

Aristotle proceeds to challenge the *endoxa* explicitly—(1) time is a movement of the whole; (2) time is a sphere; (3) time is motion and a kind of change. He readily dismisses the first two. Regarding the first view,[20] Aristotle responds that part of the *kinêsis*, or revolution, is time as much as is the whole (218b2). On the other hand, if time were the *kinêsis* of the whole and if there were more than one whole, each one revolving would be time. Aristotle waves this off as nonsensical, since this would allow for the existence of multiple times at the same time (218b4–5).

Regarding the second view that time is the sphere of the whole itself,[21] Aristotle supposes that this idea is based on the logic that (1) all things are in the sphere of

---

[20]According to Ross (1936, 596), this is a reference to Plato's *Timaeus* 37c–39e, specifically 39d1. This is also reported by Eudemus, Theophrastus, and Alexander (Simplicius 1895, 108), specifically 39C (Simplicius 1895, 111).

[21]According to Hussey (1983, 141) this is a reference to Pythagorean DK 58 B 33 or Aetius I.21, 1. According to Ross (1936, 596), Simplicius attributes it to Pythagoreans by way of a misreading of Archytas by Iamblichus "*diasteima teis tou pantos phuseos.*" According to Simplicius (1895, 108), this is something attributed the Pythagoreans, who may have misinterpreted what some Stoics reported to be Archytus's definition of time, "time was an interval in the nature of the whole." Ross clarifies that the "some Stoics" mentioned by Simplicius was Iamblichus.

the whole and (2) all things are in time (218b7). He dismisses this out of hand as naïve, and he moves on to the only theory of his predecessors that seems worthwhile to discuss.

That time is "supposed to be motion or a kind of change" (ἐπεὶ δὲ δοκεῖ μάλιστα κίνησις εἶναι καὶ μεταβολή τις ὁ χρόνος) is taken up next (218b10–11). Aristotle reasons that time is not *kinêsis* because *kinêsis* is *in* the thing that changes and *where* the thing, which moves, is (218b12–13). Time, on the contrary, is "present equally everywhere and with all things" (ὁ δὲ χρόνος ὁμοίως καὶ πανταχοῦ καὶ παρὰ πᾶσιν) (218b13–14).

Aristotle refutes the commonly held beliefs on their own terms, which is to say that he is arguing against the theories based on internal inconsistencies, assuming as his predecessors did, a concept of time as infinite time. For example, were he opposing theories of infinite time based on a theory of time taken, he would not have concluded that there cannot be simultaneous times, nor would there have been any problem with assuming that the movement of the whole would be time just as much as the movement of a part of the whole. So too, there would be no problem understanding time as equally everywhere and with all things, in contrast to motion. Change for Aristotle, recall, occurs only in terms of being; it is specific to the being undergoing the motion. Time, on the other hand, is going to end up being a number or measure—that which is not specific to a given being.

Despite that commentators have taken Aristotle's arguments in Chap. 10 so seriously as part of his analytic on time, it seems clear—when reading it as parallel to Aristotle's previous treatments of the other terms of motion—that he is here simply exposing the problems with the *endoxa* and setting himself up to re-understand time as an appropriate topic for physics. If time is to be a subject for physics, and if, as Aristotle has just shown, it is not a natural self-subsistent being itself (it defies the principles of nature previously set out), it will have to be something derived from nature. Indeed, as we have seen, Aristotle considers it a term of *kinêsis*, and he will go on to note here that it will be an attribute of *kinêsis*. In this preliminary investigation, then, he shows us only that time is not a whole composed of actual parts, which calls into question whether or not time is a continuum, but, more fundamentally, as we have seen, that time is a self-subsistent being itself.

2.3 Aristotle's Positive Account of Time

In Chap. 11, Aristotle moves on to his analytic of time. This is where he will take up the question regarding the nature of time despite that he has given his reader no good reason to think that time actually exists. This is an important point to carry over from Chap. 10. If time does not really exist, then (1) what can we really say about it, and (2) in what sense could it exist?

Aristotle introduces his analytic with what I consider to be a sort of preamble; first, he establishes time, like infinity, place, and void, to be an attribute of motion.

He begins with an argument for the coexistence of time and *kinêsis*. Time does not exist without *kinêsis*, he concludes, because it does not seem to us that time has elapsed when we have not noticed *kinêsis*. He submits the example of those fabled to sleep among the heroes of Sardinia who when awakened did not realize that any time had passed. They conflate the "now" they experience when awakened with the "now" experienced before falling asleep. Since they do not perceive the change that has in fact taken place, they fail to notice the time interval (218b21–27). Aristotle continues with an analogy—just as if the "now" were one and the same, time would not exist, when different nows are not perceived as such, it does not seem that the interval separating them is in time (218b27–29).[22] Aristotle then reasons that time is not independent of *kinêsis* (218b31), if it is true that there is no realization that time exists when there is no perception of *kinêsis*.

This is a peculiar claim because, on the one hand, Aristotle seems to be saying that time does exist independently of perception. When the difference between nows is not perceived, time is not perceived, but Aristotle seems clear here that just because time is not perceived does not mean that it does not exist. Yet, he supports his conclusion that time does not exist independently of *kinêsis* because time is not perceived without the perception of *kinêsis*; put another way, time perception entails perception of *kinêsis*. So, on the one hand, he explains time as something that exists independently of perception; and, on the other hand, he justifies this on the basis of what is perceived, i.e., on account of the inextricability of time perception with perception of *kinêsis*. These first arguments in Aristotle's analytic establish the preamble to the rest of his analytic and point to his theory of time as a time interval—a result of an interaction between a being undergoing *kinêsis* and one that is "taking" or apprehending the time of the *kinêsis*.[23] Before continuing, let us take note of the language Aristotle's argument employs here. He tells of time apprehension as a noticing, as a perceiving (218b30–35)[24]: not using the language of measure and number, as he will later on in the treatise, but the terms, ὁρίσωμεν and αἰσθώμεθα, "we mark" and "we perceive," respectively. The specific import of this passage for a full understanding of time apprehension will be dealt with in the next chapter, i.e., when I ask *who* or *what* Aristotle intends to be capable of time apprehension. For now, it is enough to notice that Aristotle's transition from

---

[22]ὥσπερ οὖν εἰ μὴ ἦν ἕτερον τὸ νῦν ἀλλὰ ταὐτὸ καὶ ἕν, οὐκ ἂν ἦν χρόνος, οὕτως καὶ ἐπεὶ λανθάνει ἕτερον ὄν, οὐ δοκεῖ εἶναι τὸ μεταξὺ χρόνος.

[23]Hussey (1983, 142) claims that, "Aristotle is arguing here from the phenomenology of time and change," which he notes to be good dialectical method and apparently "carefully non-committal" about whether time is a "content-noun" or a "mass-term." If Hussey intends the difference between "content-noun" and "mass-term" to be analogous to Aristotle's differentiation between "time taken" and "infinite time," respectively, which I suspect he does, I disagree that this ambiguity continues in Chap. 11; rather, it is relegated to Chap. 10.

[24]εἰ δὴ τὸ μὴ οἴεσθαι εἶναι χρόνον τότε συμβαίνει ἡμῖν, ὅταν μὴ ὁρίσωμεν μηδεμίαν μεταβολήν, ἀλλ' ἐν ἑνὶ καὶ ἀδιαιρέτῳ φαίνηται ἡ ψυχὴ μένειν, ὅταν δ' αἰσθώμεθα καὶ ὁρίσωμεν, τότε φαμὲν γεγονέναι χρόνον, φανερὸν ὅτι οὐκ ἔστιν ἄνευ κινήσεως καὶ μεταβολῆς.

critiquing the *endoxa* to providing his own position tells of this apprehension as the result of perception and not, perhaps, of intellection.

The analytic begins in earnest at 219a1 when Aristotle claims that, "it is evident, then, that time is neither *kinêsis* nor independent of *kinêsis*" and then announces that this will be his starting point (ἐπεὶ οὖν οὐ κίνησις, ἀνάγκη τῆς κινήσεώς τι εἶναι αὐτόν). His task now, he offers, is to understand what time has to do with *kinêsis* (219a3–4). He begins again to show that we perceive (αἰσθανόμεθα) time and *kinêsis* together. Aristotle famously concludes that there is an inextricable relationship between *kinêsis* and *chrónos* (219a4–9).[25] Though, instead of justifying the relationship based on his prior arguments that time is a term of movement, he now supports the idea based on everyday experience with time recognition. Aristotle posits that we perceive movement and time together. His evidence is again based on experience: even in the darkness, when sight is impossible or limited, and when the body does not otherwise sense change (μηδὲν διὰ τοῦ σώματος πάσχωμεν), but movement takes place in the soul (ἐν τῇ ψυχῇ) we say time has elapsed (ἅμα δοκεῖ τις γεγονέναι καὶ χρόνος). Likewise, he tells that when we think time has passed, we assume that *kinêsis* has occurred. We associate the passing of time with change and change with the passing of time. Given his previous separation of *kinêsis* from time, Aristotle immediately denies that time is actually *kinêsis*, thereby concluding that time is an attribute of *kinêsis* (ἀνάγκη τῆς κινήσεώς τι εἶναι αὐτόν). At this point, apprehension of time does not seem to require anything more than perception of *kinêsis*. Motion, in some sense, points us to time.

Aristotle then starts in another vein, establishing the relationship of *kinêsis*, thus time, with magnitude (219a10–14).[26] Since what is moved is moved from something to something and all magnitude is continuous, *kinêsis* entails the magnitude. Since *kinêsis* entails the magnitude and the magnitude is continuous, the *kinêsis* is continuous. Since *kinêsis* is continuous, time belongs to *kinêsis* (219a9), and the time that has passed is always thought to be as great as the *kinêsis*; time is, at least in some way, continuous.

Having now established the relationship of time to magnitude, Aristotle continues then to transpose the distinction of "before" and "after," one he admits to hold primarily of place and in virtue of relative position (219a15–16), to time. He moves from what he thinks must be the correspondence of "before" and "after" in place to that of *kinêsis* (219a17), and from "before" and "after" in *kinêsis* to that of time (219a18). That Aristotle argues from magnitude to time both in the case of continuity and in the case of "before" and "after" demonstrates the primacy of

---

[25]ἅμα γὰρ κινήσεως αἰσθανόμεθα καὶ χρόνου· καὶ γὰρ ἐὰν ᾖ σκότος καὶ μηδὲν διὰ τοῦ σώματος πάσχωμεν, κίνησις δέ τις ἐν τῇ ψυχῇ ἐνῇ, εὐθὺς ἅμα δοκεῖ τις γεγονέναι καὶ χρόνος. ἀλλὰ μὴν καὶ ὅταν γε χρόνος δοκῇ γεγονέναι τις, ἅμα καὶ κίνησίς τις δοκεῖ γεγονέναι. ὥστε ἤτοι κίνησις ἢ τῆς κινήσεώς τί ἐστιν ὁ χρόνος. ἐπεὶ οὖν οὐ κίνησις, ἀνάγκη τῆς κινήσεώς τι εἶναι αὐτόν.

[26]ἐπεὶ δὲ τὸ κινούμενον κινεῖται ἔκ τινος εἴς τι καὶ πᾶν μέγεθος συνεχές, ἀκολουθεῖ τῷ μεγέθει ἡ κίνησις· διὰ γὰρ τὸ τὸ μέγεθος εἶναι συνεχὲς καὶ ἡ κίνησίς ἐστιν συνεχής, διὰ δὲ τὴν κίνησιν ὁ χρόνος· ὅση γὰρ ἡ κίνησις, τοσοῦτος καὶ ὁ χρόνος αἰεὶ δοκεῖ γεγονέναι.

magnitude to time in his account (on primacy of change in place to all other *kinêsis* see also *Meta.* xii 7, 1073a10–13).

The diversion to establish the primacy of magnitude to time benefits Aristotle's account because it establishes that there is a before and after in time, but not in the circular sense in which temporality has to be assumed in order to conclude the existence of time as an attribute of *kinêsis*.[27] Instead, before and after are transposed from attributes of magnitude to attributes of time by way of the attributes of *kinêsis* to show that they constitute nothing temporal at all. Instead, they are modal, signaling the constant change of natual being from potentiality to actuality. In locomotion, due to the nature of change of place, this modality happens to be spatial. Time for Aristotle comes later. The *kinêsis* from before to after is noticed because before, "x," alters. It no longer exists; it becomes, "x_1"—after. It is thus when the *kinêsis* is noticed that time is said to have elapsed. So, while the potential for the continuity of time exists even at the same level as the continua of magnitude and *kinêsis* and, even more fundamental, the being undergoing the change, it does not exist in actuality unless the modal change from before to after is perceived. Thus, time is not the *kinêsis* from before to after.

Once Aristotle accounts for a non-temporal before and after in time, establishing these as modal features of change instead of as parts of time themselves, he turns back to his argument for the relationship between time and *kinêsis* (219a22–29)[28]:

> But we apprehend time only when we have marked motion, marking it by before and after; and it is only when we have perceived before and after in motion that we say time has elapsed. Now we mark them by judging that one thing is different from another, and that some third thing is intermediate to them. When we think of the extremes as different from the middle and the soul pronounces that the 'nows' are two, one before and one after, it is then that we say that there is time, and this that we say is time. For what is bounded by the 'now' is thought to be time—we may assume this.[29]

---

[27]My reading here has benefitted greatly from Roark's account of the "before" and "after" as non-temporal (Roark 2011, 95–119). Roark argues against the majority view that Aristotle's definition of time is circular because it uses seemingly temporal terms, i.e., "before" and "after" in the definition (Cf. Annas 1975; Owen 1975; Ross 1936 for the alternative view). But, as helpful as Roark's account is, it does not seem necessary to accept Roark's hylomorphic reading of Aristotle's Treatise on Time to understand Aristotle to intend an underlying material continuum to provide non-temporal "relata" expressed in the relation "before" and "after." Roark argues that priority and posteriority are already present in Aristotle's account of *kinêsis* (Roark 2011, 95). I agree, but they are present only insofar as there is a natural being undergoing *kinêsis*.

[28]ἀλλὰ μὴν καὶ τὸν χρόνον γε γνωρίζομεν, ὅταν ὁρίσωμεν τὴν κίνησιν, τὸ πρότερον καὶ ὕστερον ὁρίζοντες· καὶ τότε φαμὲν γεγονέναι χρόνον, ὅταν τοῦ προτέρου καὶ ὑστέρου ἐν τῇ κινήσει αἴσθησιν λάβωμεν. ὁρίζομεν δὲ τῷ ἄλλο καὶ ἄλλο ὑπολαβεῖν αὐτὰ καὶ μεταξύ τι αὐτῶν ἕτερον· ὅταν γὰρ ἕτερα τὰ ἄκρα τοῦ μέσου νοήσωμεν, καὶ δύο εἴπῃ ἡ ψυχὴ τὰ νῦν, τὸ μὲν πρότερον τὸ δ' ὕστερον, τότε καὶ τοῦτό φαμεν εἶναι χρόνον· τὸ γὰρ ὁριζόμενον τῷ νῦν χρόνος εἶναι δοκεῖ· καὶ ὑποκείσθω.

[29]Hardie and Gaye translate ἡ ψυχὴ, "mind" in the ROT. To be more precise, I have amended the translation so that ἡ ψυχὴ is rendered "soul."

Here again, we see Aristotle allow that the apprehension of time requires *marking* change (τὸν χρόνον γε γνωρίζομεν ὅταν ὁρ σωμεν τὴν κίνησιν).[30] And again, it is not change per se that is marked, but the "before" and "after." This before and after refers to the "one thing...different from another", i.e., of the differing modal features of a being undergoing *kinêsis* in the context of its underlying unity, viz. that it is a self subsistent substantial being *qua* itself. When we apprehend the difference between modal features in terms of a substantial being undergoing various sorts of change (alteration, locomotion, diminution/growth), we ascertain that the nows are more than one.[31]

This is to say that time, meaning the time taken, appears to exist because it is apprehended by us as a result of (1) the principle of nature, and (2) the apprehension of that principle. "Now," which is terminology precipitate of the *endoxa*, takes on a modal reference. "Now," recall, is not an actual part of time (see also ahead at 220a18–21); it is merely believed to be a part of time. It is a limit. It delimits the *kinêsis* occurring of existing self-subsistent beings (Recall 211b30–212a2 from Aristotle's discussion of place. He related place first to the *hypokeimenon*, or the intermediate that undergoes change, and then as a limit). The man is untrained (now α), and he is trained (now β). When accidental change is noticed and marked, time is said to have elapsed, at least in some sense.

Aristotle continues, again drawing conclusions about what time *is* with support from the way it is perceived (recall 218b27–29). Time is not thought to have elapsed, he reasons, when the "now" is not perceived to be more than one (219a30–31). But, whereas in the previous argument Aristotle leaves open the possibility that time exists regardless of the perception, and that it is only on account of the perception— or lack thereof—when we misapprehend time, here Aristotle makes the stronger claim that the actual perception and subsequent apprehension of the before and after contributes necessarily to the being of time. "When we do perceive a 'before' and an 'after,'" he writes, "then we say that there is time. For time is just this—number of motion in respect of 'before' and 'after'" (ὅταν δὲ τὸ πρότερον καὶ ὕστερον, τότε λέγομεν χρόνον· τοῦτο γάρ ἐστιν ὁ χρόνος, ἀριθμὸς κινήσεως κατὰ τὸ πρότερον

---

[30]Heinemann's analysis of this passage (2012, 6) is helpful here: "The 'earlier and later' in change is that, by being which in passing change is. Yet, what it is to be earlier and later in change is something else, and is not the same thing as change. We become acquainted with time when we mark off the change, that is, when we mark it off by what is earlier and later. We say that time has passed when we get a perception of the earlier and later in change."

[31]Looking at language used: Aristotle is here referring to thinking or judging for the first time, Aristotle tells that when we judge a difference between *this* (thing here) "now" and *this* (thing here) "now," we mark time. However, it is the soul (ἡ ψυχὴ), not specified to be either perceptive or intellective, which discriminates the nows—the before and after—as two. *De motu* 700b18–20 uses different language (κριτικὰ instead of νοήσωμεν) to conclude that perception is capable of exercising judgment: "For imagination and sensation cover the same ground as the mind (since they all exercise judgment) (though they differ in certain aspects as has been defined elsewhere" (*De anima* iii). This early language about perception, marking, and apprehension of time by soul, generally, suggests that time may be apprehended by the sensitive soul as well. Again, we will return to this possibility in the next chapter.

καὶ ὕστερον)^{32} (219a34–219b1). Not only do we say "there is time," when we perceive the change from "before" to "after," but this time that we proclaim when we have apprehended it is indeed all that time is if we are talking about the time taken. Just prior to his famous definition of time, Aristotle concludes: "for time is this" (τοῦτο γάρ ἐστιν ὁ χρόνος). His original puzzle to understand the being of time has properly debunked the *endoxa*, and in their wake leaves an entirely new way to think about temporality. Not unlike his treatments of place, infinity, and void, we see his clear intention here to associate time with *kinêsis* and, more primordially, with the beings undergoing *kinêsis,* to render time a potential derivation of *kinêsis* when certain conditions are met.

Predictably, then, Aristotle again concludes that "time is not *kinêsis*," and here he adds the clarification that it is, "only *kinêsis* in so far as it admits of enumeration" (219b3–5). At this point, then, Aristotle has moved through his justification for the conclusion that time does exist despite that it seems impossible that it could. It exists because Aristotle has redefined it. Now, time is to be understood as a number and not as an imagined vessel containing parts that do not exist. It is to be understood as a number, which demarcates each interval of *kinêsis* for natural beings when this *kinêsis* is apprehended. To say that time exists, then, is to significantly qualify what "exists" means. This is where we have to rely once again on the modal category of beings that exists only in potentiality, which Aristotle established in *Physics* iii.

Aristotle posits a substantial natural being, a subject or a "this," to help demonstrate what he means by "now." The particular example of primary substance he posits is a human being (τὸ Κορίσκον) (219b13–33), which is meant to be a metaphor for the "now." The idea that the now is a being *itself*, and a perfect being at that, is dismissed. The human or "this" is carried from place to place. As the "this" travels, time seems to progress. So is the "this" the same "this" in each place? Or does the "this" change? Aristotle's solution is two-fold. It reiterates our earlier point about the way that natural self-subsistent beings undergo *kinêsis*. On the one hand, the "this" stays the same because there is something about the "this" that does not substantially changes as it moves along. In order for *kinêsis* to even exist, there must be something that is undergoing the *kinêsis*. Aristotle calls this aspect of something its substratum. In the case of the human being, there is an underlying unity of material and form unmarred as the "this" is carried along. Its identity remains intact not only in its starting location, but also in each location where it arrives thereafter. On the other hand, the "this" changes or moves in accordance with its various potentialities for accidental change (see again 192b13–22). With its travels, we can imagine it ages in accordance with the succession of its locomotion; it is altered in small—even superficial—ways, e.g., it may become pale, thinner, weaker. The "this" both remains the same and yet is ever different. 219a22–29.

---

^{32}The "we perceive" given here in English but not found in the Greek is a carryover from the "we perceive" αἰσθανώμεθα just previous in 219a31; the two clauses are parallel in sentence structure.

What Aristotle means here is that the now, like each "this," is non-temporal; it is something that exists and changes along the continuum of spatial magnitude, itself a modal continuum in the sense that it is ever changing from potentiality to actuality.[33] The body is "here" and now it is "there." In this case, where there is an explicit display of change in place, the change in the "this" is noticed as a spatial difference; it can be moved in any direction—it is not necessarily moved in the typical forward processing temporal direction "left" to "right." Despite its direction, its change from "here" to "there" is perceived and marked. Recall 219a22–29, the "this" is a foundation primary to the "now."

Just as we become aware of "before" and "after" in the act of the subject being carried, yet despite the direction it is moving, we likewise notice the "now" when we observe the "this" undergoing change, whether in terms of place or in terms of qualitative or quantitative change.[34] "Before" the alteration is differentiated from "after" the alteration because a change is perceived. The house was not built, and now it is built; I was on my way to Thessaloniki, and now I have arrived. We typically think of these examples of *kinêsis* as temporally determined. We understand time to be a vessel in which all change occurs according to a predetermined progression, and we think of the "now" as points on the line of this progress. But, this view is precisely what Aristotle has countered. The "now," as it is with the human subject Aristotle posits, is both that which remains, i.e., the identity or substratum that is maintained through *kinêsis*, as well as the difference before and after the *kinêsis* (219b26–29). The "now" is every subsistent being, both its substratum and its difference between what it is before and then after *kinêsis*.

Recall, that Aristotle is after comprehensive understanding of nature, and here he writes, "this is what is most knowable; for motion is known because of that which is moved, locomotion because of that which is carried. "For what is carried is a 'this' (τόδε τι), the movement is not" (219b29-31). *Physics* i–ii provided us the *archai* of nature and the nature of natural beings. *Physics* iii–iv investigated motion and its terms. Here, we see confirmation from Aristotle that we have indeed been

---

[33]Coope (2005, 29) supports that the now is not temporal when she observes, "On the one hand, none of time *is* except the now. This suggests that time only exists in virtue of the existence of the now. But on the other hand, for the now to exist, it must be a division or boundary of some independently existing continuum. This continuum cannot be *time*, since time itself is dependent on the now. It follows that there must be some *other* continuum, prior to time, on which the now depends for its existence" (emphases in original). For Coope, however, the "other continuum" is going to be change. I will ultimately disagree with this conclusion. The more primordial "other continuum" is a "this," the self-subsistent existing natural beings undergoing the change as a result of their very nature. King (2009, 63) states both that "the change is marked by our saying *now* and *now*; that is how we mark off the before and after in time"; and, "saying now has to be thought of as occupying no time, like an instant…[the now] is the temporal analogue of a point…" While it is not correct to say that the before and after is marked *in* time—*by* time is more appropriate— because as King acknowledges just after, the now does not occupy time, it does seem right that the change is marked by apprehension of more than one "now," which must be non-temporal.

[34]See Hussey (1983, 143) on "changes 'along' magnitudes"; there, he concludes that every change is necessarily a change along a path and thus that there is ontological and logical priority on the path.

proceeding from what is most knowable to us to what is most knowable to nature. We perceive motion, which alerts us to investigate nature. When we investigate the nature of natural beings, we find that their nature is the principle of *kinêsis* and stasis.

Kinêsis exposes the complexity of natural beings, humans included; no natural being, by its nature, is simply static. We proceed from the *kinêsis* we perceive, and we discover that the terms of motion are all—at least to begin with—potentialities and not actualities of being. e.g., contra Zeno, infinity exists only by potential division. The *kinêsis* itself is not the topic of investigation; the "this," or substantial beings are. The "now" we notice as "before" in this way and "after" in that way is precisely Aristotle's topic in the *Physics*, as demonstrated in the last chapter. The kind of being, which remains the same, and yet constantly changes, is peculiar to natural being. This is to say that "the now" is a common name for natural being, and thus a referent for its various stages of potentiality and actuality.

Aristotle has thus done the work to extricate the temporal character of "now" (*nun*) from the term. To perceive a change from "now" to "now" connotes no change "in time." Instead, it means simply the actual difference on the path (to use Hussey's term) of *kinêsis* from "before" (Jackson is untrained.) to "after" (Jackson is trained.). Aristotle's moving body metaphor is perfect here—the body was "here" and now it is "there." The temporal component of such *kinêsis* comes as a derivative of the *kinêsis* when the change is apprehended. It is this apprehension, perception and marking, which creates time by way of bringing it from potentiality (possible as a derivative of the change naturally occurring in this world) to actuality (actually derived of the change naturally occurring in the world by another part of nature).

With this said, then, we are in a position to correctly interpret Aristotle's subsequent claim that "if there were no time, there would be no 'now', and vice versa" (φανερὸν δὲ καὶ ὅτι εἴτε χρόνος μὴ εἴη, τὸ νῦν οὐκ ἂν εἴη, εἴτε τὸ νῦν μὴ εἴη, χρόνος οὐκ ἂν εἴη) (220a1). It would be too easy to read this passage to suggest that Aristotle has now contradicted himself, or that my argument is severely flawed, understanding him here to be reverting to a traditional understanding of time as a whole composed of three parts: past, present, and future. And, this would seem to make sense. How could we have time without having "now"? But, what Aristotle seems to mean here is that to speak of "now" as a common name for an existing self-subsistent natural being undergoing *kinêsis* is already to be implying perception of the being. Just like the number that Aristotle claims to be time, the "now" refers to or names the natural being existing independently of all perception and conceptualization. The "now" does not exist without time and vice versa because both the "now" and time require someone noticing and naming, i.e., apprehending, *kinêsis* in natural objects. Put another way, "now," signifies a relation between the one perceiving motion and the motion itself; it is a referent to mark perceived change from "before" to "after." Re-invoking the body metaphor, Aristotle concludes that, "the number of the locomotion is time, while the 'now' is comparable to (ὡς τὸ) the moving body, and is like the unit of number" (χρόνος μὲν γὰρ ὁ τῆς φορᾶς ἀριθμός, τὸ νῦν δὲ ὡς τὸ φερόμενον, οἷον μονὰς ἀριθμοῦ) (220a4). The number is the name of the change, and the now is the name of the "this"—the existing self-subsistent natural being—observed. That both the thing changing and the change itself are named implies someone or something doing the naming.

The "now" and time have a complex relationship because not only does time seem to be made continuous by the now, i.e., time intervals continue so long as a natural object is in motion, but also time is limited by the "now," i.e., when change has occurred, the interval numbering the change likewise ends (220a5). To say here that the "now" is both that which makes time continuous as well as that which limits time is really to equivocate on the term. Or, to be charitable to Aristotle here, it is seemingly to conflate the two senses of "now" just established—(1) the substratum of the natural object and (2) the object "before" and then "after" *kinêsis*. It is by the first sense of "now" that time is made continuous because the natural object continues to move with periods of rest so long as it exists. It is by the second sense that time is limited.

Aristotle returns to the earlier comparison of the "now" with a point (recall 218a12–29), officially dismissing it here (220a9–14). Whereas a point can be the end of one thing and the beginning of another, essentially making one into two, so long as there is a pause, the "now" taken in the first sense above is the analogue or name of the body constantly moving. It continuously undergoes many individual instances of *kinêsis*. Thus, it is in this sense always different. It is constantly undergoing *kinêsis* just as the body is always being carried along.

Aristotle concludes Chap. 11 asserting that the "now" in indeed not time. It is an attribute of time (ἣ μὲν οὖν πέρας τὸ νῦν, οὐ χρόνος, ἀλλὰ συμβέβηκεν). To clarify, though, Aristotle does not intend attribute (συμβέβηκεν, literally "comes together") here in the sense that time is "an attribute" of *kinêsis*, i.e., derivative of it. The sense in which the "now" is an attribute of time is "in so far as it numbers, it is number... but number (e.g., ten) is the number of these horses, and belongs also elsewhere" (220a18–21).[35] This is the first time Aristotle will introduce the Greek idea that number is nothing symbolic, but rather that which is named by the number (see also 220b6–9), i.e., "the number of these horses". Because "now" names the natural object or "this," and the "this" is constantly undergoing *kinêsis*, the number of its *kinêsis* from "here" to "there," from "before" to "after," ends up referring to the same thing, though in a different sense, that the "now" names. Number names the things counted, i.e., the "nows," and the now names, at least in one sense, the natural being at different points of *kinêsis*, i.e., at different points of being.

Following the discussion of the relationship between time and "now," Aristotle concludes *Physics* iv 11 confidently, saying: "It is clear, then, that time is number of *kinêsis* in respect of the before and after, and is continuous since it is an attribute of what is continuous" (ὅτι μὲν τοίνυν ὁ χρόνος ἀριθμός ἐστιν κινήσεως κατὰ τὸ

---

[35]ᾗ δ᾽ ἀριθμεῖ, ἀριθμός †· τὰ μὲν γὰρ πέρατα ἐκείνου μόνον ἐστὶν οὗ ἐστιν πέρατα, ὁ δ᾽ ἀριθμὸς ὁ τῶνδε τῶν ἵππων, ἡ δεκάς, καὶ ἄλλοθι.

πρότερον καὶ ὕστερον, καὶ συνεχής (συνεχοῦς γάρ), φανερόν) (220a25–26).[36] Aristotle thus ends the chapter as if he were providing a conclusion immediately following his discussion of the magnitude-*kinêsis*-time relationship at 219a14. Strangely, this abrupt back-step to what he had discussed prior to his arguments for the relationship between the "now" and time make the latter seem as though they were tangential. Perhaps Aristotle wanted to reconcile his definition of time with previous conceptions of the now; if his entire analytic of time would contend with the *endoxa*, he had to explain too a new way to think about "now," i.e., as non-temporal. If "now" is non-temporal, then so too are "before" and "after," and thus there is no circularity in his definition of time, as the number of before and after with regard to *kinêsis*. And, in this last assertion, he brings everything together when he returns to the idea that temporality is an attribute of that which is already continuous, i.e., *kinêsis*, and by way of his discussion of the now, it seems clear that *kinêsis* is in turn consequent of that which is more primordial to change, i.e., the natural being that undergoes the *kinêsis*.

If Aristotle has then addressed the first puzzle in his analytic of time and has established that time *does* exist, but in a new sense, i.e., as a potential continuum derived from the *kinêsis* beings are undergoing, it is still left to him to be more explicit about its nature. If time needs to be apprehended in order that it exist as actualized, i.e., as a number identifying the *kinêsis* of a being from before to after, who or what exactly is doing the apprehending? Whence does the number come? We will take up these questions in the next chapter.

---

[36]Hardie and Gaye (ROT) render "συνεχής (συνεχοῦς γάρ), φανερόν" as "attribute of what is continuous," but the idea of "attribute" does not appear in the Greek. It would be more accurate to translate the Greek: manifestly continuous; for the continuous.

Chapter 3
Taking Time

Despite the language we saw in the previous chapter, which allowed for time apprehension by perception and marking, in *Physics* iv 14, Aristotle famously argues that time is dependent on *nous* (see 223a25–26, ἢ ψυχὴ καὶ ψυχῆς νοῦς). In what sense could the number of motion with respect to before and after be dependent on *nous*? Because Aristotle famously discusses the relationship between time and the soul, and only once qualifies soul as *nous*, it has been common for readers to underdetermine *nous*, as simply "soul" across the treatise. This is problematic because, for Aristotle, there are five main potencies of soul: intellective, locomotive, desiring, sensitive, and nutritive. While he argues that human beings have all five, he also tells us that nonhuman animals have at least desiring, sensitive and nutritive potencies—usually they also have locomotive—and still plants have the nutritive potency (see *De anima* ii 3). If *nous* can be collapsed into meaning simply, soul, the implication is that time is in every case dependent on ensouled being generally. The term *nous*, often translated "mind" and not "soul," is problematic without the added confusion that comes from conflating it with "soul." Namely, it is both the term Aristotle uses to single out the intellective faculty of soul, which as noted is reserved for human beings, and the term often understood by Aristotle's readers as that which names God/the first principle and the celestial bodies. In order to follow Aristotle's definition of *chrónos*, it is necessary to understand how he is using *nous* in *Physics* iv 14.

In this chapter, therefore, I consider the meaning of *nous* in Aristotle's account of time as well as the necessity of a body by which to sense-perceive that which can be counted (recall 219a4–6, we perceive, αἰσθανόμεθα, time with *kinêsis*, and, by extrapolation, we perceive *kinêsis* through bodily senses, σώματος πάσχωμεν, and in the soul, ἐν τῇ ψυχῇ). As a result of my considerations, I argue for the possibility for limited time apprehension of nonhuman animals, and for the possibility for full time apprehension in human beings. This final task is buttressed both by discussions of sensation, memory, and animal behavior in the *Parva Naturalia* and in Aristotle's biological treatises and the bringing together the language of "marking time" and "counting/measuring time" to suggest that even while big-scale change

© The Author(s) 2015
C.C. Harry, *Chronos in Aristotle's Physics*,
SpringerBriefs in Philosophy, DOI 10.1007/978-3-319-17834-9_3

requires the latter, the former is sufficient for small-scale change.[1] This is all in service to showing internal consistency in Aristotle's account of time apprehension in the physical works. In the case where Aristotle's examples of animal behavior limit the search for supporting evidence, I turn to results from contemporary experimental science to show that the position I attribute Aristotle is consistent *otherwise*, with demonstrated animal—both human and nonhuman—behavior and function. I will proceed with these tasks in Sects. 3.2 and 3.3. But, first, in Sect. 3.1, I will provide a negative account and say more about what I think Aristotle does not mean when he uses *nous* in *Physics* iv 14.

3.1 Conditions for Actualized Time

In Greek, there is a way in which *nous* means not only "mind," but also "perception." It is dubious that Aristotle intends a generally unconventional use of *"nous"* in *Physics* iv 14, i.e., meaning "perception" and not "mind," as he seems to do elsewhere (see *Nic. Ethics* vi); rather, *"nous"* here means broadly the working together of sense and intellection in that, as we see in *De anima*, the faculties of intellect require sensation. This is important to Aristotle's definition of time in particular because actualizing time, in the majority of cases, requires not only perception of *kinêsis*, but also counting *kinêsis*. The being undergoing *kinêsis* does so irrespective of the apprehension. But, only a being that can both perceive and count can interact with the being undergoing *kinêsis* in such a way so as to actualize time. Super human beings have neither a faculty (*dunamis*) by which to apprehend *kinêsis*, nor the type of intellect with the potential for counting. Sub-human beings do not have a rational soul with which to count. Aristotle thus could not have meant either that actualized time depends on, on the one hand, a super human being like God or the celestial bodies, or, on the other hand, a sub-human being like non-human animals or plants.

The unmoved mover/God is neither in time, nor does God have the potential for change. Thus, some have refuted the traditional reading of *Metaphysics* xii 7 where *nous* is thought to refer to God. Instead, a distinction has been made between *nous*, which is a readiness for thinking (see *De anima* iii 4) and *noesis*, or, thinking itself. It has been argued, thus, that God is not *nous* for Aristotle, as that contradicts the idea that God is pure actuality outside of time, but *noesis* (see Polansky 2011). Further, it has been claimed that God cannot be *noesis* either for Aristotle, since even the act of thinking seems to suppose an element of potentiality in that it requires an object (*noeta*) (see Gabriel 2009). Both have important implications for understanding Aristotle's account of time, and I agree with the general thrust of both. On the first account, *nous* cannot mean God in *Physics* iv 14 because that would require God, or pure activity, to have the potentiality to number, or count, the

---

"before" and "after" in a being undergoing *kinêsis*. Ironically, this would render God impotent, since he would share the same lack of knowledge that humans, nonhuman animals, and plants have. On the second account, there is even more to find objectionable, i.e., not only is God's mind reduced to mere readiness for counting, but also it has an object of its activity, i.e., the *arithmos* of the *kinêsis*.

Aristotle begins *Metaphysics* xii 7 recounting his conclusions from *Physics* viii, that there are eternal heavens set into motion by what must be an unmoved mover. He likens the unmoved mover to objects of thought and desire; they too move without being moved (1072a26–27). Aristotle then demonstrates that whatever cannot be moved also cannot be that which is moved by an object of thought (1072a26–1072b1).[2] This passage differentiates *noesis* (thinking), *nous* (readiness, i.e., a potentiality, for thinking) and *noeta* (object of thought). The term in question is *nous*, which according to this passage has the capacity to receive objects of thought—a capacity that the unmoved mover could not have—not least of all because that which only "exists actually" has no capacity, i.e., potentiality at all. Consider, for example, that the unmoved mover, as the first mover, is not only the first in its class, but by virtue of this, the best. If the unmoved mover is the best object of thought, it is clearly an object of thought. Objects of thought move thought. Yet, it is impossible that the unmoved mover move itself. The unmoved mover does not have motion. If the unmoved mover is an object of thought, it is thus not also moved by thought.

Further, Aristotle explains, since thought shares the nature of the object of thought, readiness for thinking can think itself. Thought and object of thought can be the same thing (αὐτὸν δὲ νοεῖ ὁ νοῦς κατὰ μετάληψιν τοῦ νοητοῦ) (1072b20–21). But, again, thought (νοῦς) here cannot refer to the unmoved mover/God. God has no capacity to think itself. Thinking for Aristotle, when human thinking, is not an isolated activity of an intellective capacity; rather, it occurs as a relation between a rational soul (*nous*) who has the capacity for receiving an object of thought, i.e., perception, and the readiness to think about it, i.e., intellection. In order to be both that which is thinking and that which is the object of thought, something must have the potential for actual thinking. *Nous* here refers instead to the intellective faculty of the soul. That the rational soul can make an object of itself shows that the rational soul is a potentiality of an existing self-subsistent being, who is itself a natural being. While the actuality of the divine is something toward which *nous* always strives, it is the potentiality of *nous* and of all natural objects, which characterizes them as existing self-sufficient natural beings.

---

[2] κινεῖ δὲ ὧδε τὸ ὀρεκτὸν καὶ τὸ νοητόν: κινεῖ οὐ κινούμενα. τούτων τὰ πρῶτα τὰ αὐτά. ἐπιθυμητὸν μὲν γὰρ τὸ φαινόμενον καλόν, βουλητὸν δὲ πρῶτον τὸ ὂν καλόν: ὀρεγόμεθα δὲ διότι δοκεῖ μᾶλλον ἢ δοκεῖ διότι ὀρεγόμεθα: ἀρχὴ γὰρ ἡ νόησις. νοῦς δὲ ὑπὸ τοῦ νοητοῦ κινεῖται, νοητὴ δὲ ἡ ἑτέρα συστοιχία καθ' αὑτήν: καὶ ταύτης ἡ οὐσία πρώτη, καὶ ταύτης ἡ ἁπλῆ καὶ κατ' ἐνέργειαν (ἔστι δὲ τὸ ἓν καὶ τὸ ἁπλοῦν οὐ τὸ αὐτό: τὸ μὲν γὰρ ἓν μέτρον σημαίνει, τὸ δὲ ἁπλοῦν πῶς ἔχον αὐτό). ἀλλὰ μὴν καὶ τὸ καλὸν καὶ τὸ δι' αὐτὸ αἱρετὸν ἐν τῇ αὐτῇ συστοιχίᾳ: καὶ ἔστιν ἄριστον ἀεὶ ἢ ἀνάλογον τὸ πρῶτον.

For Aristotle, actual rational thought depends on the potentiality (*dunamis*) for thought, and this is consequent on the capacity (*dunamis*) to receive the object of thought (1072b21–22). The thinking is actual, which is to say it is in the process of thinking, when it possesses the object (1072b22). It is this active element, which Aristotle calls, God-like (δοκεῖ ὁ νοῦς θεῖον ἔχειν). Aristotle next argues that God's nature is essentially different from the nature of existing self-subsistent natural beings (1072b24–30).[3] Some natural beings are "God-like" in that they have a rational soul; for Aristotle, these are human beings. God is eternal, whereas humans are mortal, God is superlative, whereas humans share in a piece of God's goodness, God is actuality and life, whereas substantial beings are by nature ever potentially other than what they are now; their nature is the potential for *kinêsis*. Aristotle turns to the nature of divine thought in *Metaphysics* xii 9, concluding there that God or "God's thinking" is "thinking on thinking" (ἔστιν ἡ νόησις νοήσεως νόησις) (1074b34). In other words, God is pure actuality (*energeia*). Recall that the inherent potentiality for *kinêsis* (and likewise, rest) is the nature of natural beings. The first mover/God is pure actuality and cannot be otherwise; hence, it is not capable of *kinêsis*. Since pure actuality has no readiness to receive perceptibles, and in fact no potentiality whatsoever in its being, it cannot carry out the functions requisite to apprehend or take time.

A reading of *De memoria et reminiscentia* indicates that nonhuman animals experience time. Aristotle begins the treatise announcing that he will now treat memory and remembering. He will consider not only its nature and its cause, but also the part of the soul to which these functions, along with recollecting, belong (449b4–6). The distinction made here between memory and recollecting is important for Aristotle; for example, he goes on to clarify that the former is generally sharper in slow people, while the latter is generally sharper in clever people (449b7–8). The objects of memory, he argues, are relegated completely to things that are past (449b14). The future is not remembered, but expected, and the present is sense perceived (449b10–13). Aristotle demonstrates this to be the case with an example. When one is sensing a white object before him, he would say he is perceiving it, not remembering it. Likewise, when one is contemplating an object of science in a given moment, he would say that he knows it, not that he is remembering it.

When the objects are not being perceived or thought readily, then they are being remembered. Remembering, for Aristotle, reconstitutes previously learned knowledge or previous sense perception in one's mind (449b15–24). It brings to mind an activity that has since ceased. He concludes that, "memory is, therefore, neither perception nor conception (ὑπόληψις), but a habit or state of one of these, whenever time has become (ἕξις ἢ πάθος, ὅταν γένηται χρόνος)" (449b25).[4]

---

[3]εἰ οὖν οὕτως εὖ ἔχει, ὡς ἡμεῖς ποτέ, ὁ θεὸς ἀεί, θαυμαστόν: εἰ δὲ μᾶλλον, ἔτι θαυμασιώτερον. ἔχει δὲ ὧδε. καὶ ζωὴ δέ γε ὑπάρχει: ἡ γὰρ νοῦ ἐνέργεια ζωή, ἐκεῖνος δὲ ἡ ἐνέργεια: ἐνέργεια δὲ ἡ καθ᾽ αὑτὴν ἐκείνου ζωὴ ἀρίστη καὶ ἀίδιος. φαμὲν δὴ τὸν θεὸν εἶναι ζῷον ἀίδιον ἄριστον, ὥστε ζωὴ καὶ αἰὼν συνεχὴς καὶ ἀίδιος ὑπάρχει τῷ θεῷ: τοῦτο γὰρ ὁ θεός.

[4]Beare translates ἕξις, "affection," seemingly missing the ambiguity of the term, i.e., that it might mean habit or potentiality/disposition. He renders ὅταν γένηται κρόνος, "conditioned by a lapse of time" in the ROT.

The consequence of Aristotle's definition of memory is that, "only those animals which perceive time remember, and the organ whereby they perceive time is also that whereby they remember" (ὥσθ᾽ ὅσα χρόνου αἰσθάνεται, ταῦτα μόνα τῶν ζῴων μνημονεύει, καὶ τούτῳ ᾧ αἰσθάνεται) (449b29–30). Thus, on Aristotle's account, time perception (κρόνου αἰσθάνεται), which implies the ability either for sense perception or intellection, or for both, is the necessary and sufficient condition for memory. We must determine the organ by or through which time perception happens, then, in order that we understand the types of animals that perceive time. Deciding the organ by or through which time perception happens may also be additional evidence that we can rule out God as a sufficient condition for the actuality of time, since as we have seen, God does not have parts, thus cannot have organs for time perception.

Aristotle appeals to his argument from *De anima* regarding the necessity of images for thinking (449b31–450a8).[5] Aristotle posits subsequently that, "we cannot think of anything without a continuum or think of non-temporal things without time" (450a9–10), a fascinating admission to which he does not return. It is possible that Aristotle is referencing his claim from *Physics* iv 12 that things not measured are not necessarily "in time," but only accidentally in time (221b25). Even if non-temporal, which I imagine entails not undergoing *kinêsis*, Aristotle imagines that something can be accidentally "in time" insofar as it exists in concert with things that are undergoing *kinêsis* and being measured. Next, Aristotle builds on his previous argument, now showing that thought and thinking are only incidental to memory (450a9–14).[6] The sense in which intellection is only incidental to sense perception in the case of memory is that intellection depends on sense perception, even remotely in the case of intellectual objects since it is impossible to think without having had any experience at all with sense perception. Thus, Aristotle is saying here that there is the possibility for memory, which requires only the faculty of sense perception. Whereas, memory can be aided by intellection derived from sense experience, this is not a necessary condition for memory. This reasoning allows Aristotle then to conclude that, "Hence not only human beings and the beings which possess opinion or intelligence, but also certain other animals, possess memory" (διὸ καὶ ἑτέροις τισὶν ὑπάρχει τῶν ζῴων, καὶ οὐ μόνον ἀνθρώποις

---

[5]νοεῖν οὐκ ἔστιν ἄνευ φαντάσματος· συμβαίνει γὰρ τὸ αὐτὸ πάθος ἐν τῷ νοεῖν ὅπερ καὶ ἐν τῷ διαγράφειν· ἐκεῖ τε γὰρ οὐθὲν προσχρώμενοι τῷ τὸ ποσὸν ὡρισμένον εἶναι τοῦ τριγώνου, ὅμως γράφομεν ὡρισμένον κατὰ τὸ ποσόν, καὶ ὁ νοῶν ὡσαύτως, κἂν μὴ ποσὸν νοῇ, τίθεται πρὸ ὀμμάτων ποσόν, νοεῖ δ᾽ οὐχ ᾗ ποσόν· ἂν δ᾽ ἡ φύσις ᾖ τῶν ποσῶν, ἀορίστων δέ, τίθεται μὲν ποσὸν ὡρισμένον, νοεῖ δ᾽ ᾗ ποσὸν μόνον.

[6]μέγεθος δ᾽ ἀναγκαῖον γνωρίζειν καὶ κίνησιν ᾧ καὶ χρόνον· [καὶ τὸ φάντασμα τῆς κοινῆς αἰσθήσεως πάθος ἐστίν] ὥστε φανερὸν ὅτι τῷ πρώτῳ αἰσθητικῷ τούτων ἡ γνῶσίς ἐστιν· ἡ δὲ μνήμη, καὶ ἡ τῶν νοητῶν, οὐκ ἄνευ φαντάσματός ἐστιν, <καὶ τὸ φάντασμα τῆς κοινῆς αἰσθήσεως πάθος ἐστίν>· ὥστε τοῦ νοῦ μὲν κατὰ συμβεβηκὸς ἂν εἴη, καθ᾽ αὑτὸ δὲ τοῦ πρώτου αἰσθητικοῦ.

καὶ τοῖς ἔχουσι δόξαν ἢ φρόνησιν) (450a14–15). When we connect this conclusion with the prior claim that animals that sense time also have memory, we are tempted to conclude that nonhuman animals, insofar as they have the faculty of sense perception, perceive time. When we consider *Physics* iv 14, we see that this can not be the whole story. Though it is clear that Aristotle intends that, in some sense, time is perceived, there must be a limit to this kind of time apprehension in order that *Physics* iv 14 be consistent with Aristotle's remarks elsewhere.

Aristotle next clarifies that memory entails apprehension of before and after (450a19–20), which one assumes if memory entails time sense and if time sense entails apprehension of before and after. He then gets specific when he writes, "if asked, of which among the parts of the soul memory is a function, we reply: manifestly of that part to which imagination also pertains" (τίνος μὲν οὖν τῶν τῆς ψυχῆς ἐστι μνήμη, φανερόν, ὅτι οὗπερ καὶ ἡ φαντασία) (450a21–22). Aquinas, in his commentary on *De memoria et reminiscentia*, explains that apprehension of before and after entails imagination (*phantasia*):

> For some animals perceive nothing save at the presence of sense objects, such as certain immobile animals, which on this account have an indeterminate imagination, as *De anima* iii says. And on this account they cannot have cognition of prior and posterior, and consequently nor time. Hence they do not have memory.

It is not simply animals with sense perception that have memory, but animals with the ability to determine that "this" perceptible is being perceived "before" or "after" "this" perceptible. This determination requires an ability to mark (*orizei*) *kinêsis* in some sense. Here we find an indication that even if some or many nonhuman animals perceive time, not all can—owing to a lack of determinate imagination.

Aristotle ends the first chapter writing, "it has been shown that it [memory] is a function of the primary faculty of sense perception, i.e., of that faculty whereby we perceive time (ὅτι τοῦ πρώτου αἰσθητικοῦ, καὶ ᾧ χρόνου αἰσθανόμεθα)" (451a16–17). That time is *perceived* (ᾧ χρόνου αἰσθανόμεθα) by the faculty of *sense perception*—for Aristotle this is a faculty of the sensitive soul, and thus perceived by any being endowed with sense—seems unproblematic. In fact, this language is perfectly consistent with what Aristotle tells us about time perception in *Physics* iv 11. The problem is with Aristotle's argument at *Physics* iv 14, to which we will now turn.

3.2 Readiness for Thinking: From Marking to Counting

Aristotle's discussion of the dependence of time on the soul is one of the more famous and debated passages in the time section of the *Physics*. Despite its relative brevity—spanning a mere paragraph of the overall argument—interpreters have disagreed about how to understand the crucial relationship Aristotle posits among time, *arithmos*, soul, and *nous*. The passage reads as follows (223a16–28)[7]:

> It is also worth considering how time can be related to the soul; and why time is thought to be in everything, both in earth and in sea and in heaven. Is because it is an attribute, or state, or movement (since it is the number of movement) and all these things are movable (for they are all in place), and time and movement are together, both in respect of potentiality and in respect of actuality? Whether if soul did not exist time would exist or not, is a question that may fairly be asked; for if there cannot be some one to count there cannot be anything that can be counted, so that evidently there cannot be number; for number is either what has been, or what can be, counted. But if nothing but soul, or in soul reason, is qualified to count, there would not be time unless there were soul, but only that of which time is an attribute, i.e., if movement can exist without soul, and the before and after are attributes of movement, and time is these qua numerable.

Let us begin with a general observation. Notice here that Aristotle is recalling his actual definition of time from *Physics* iv 11, talking about time as a number, *arithmos*. And here, he takes a step further to define number. Number is something that has been or can be counted: Ἀριθμὸς γὰρ ἢ τὸ ἠριθμημένον ἢ τὸ ἀριθμητόν. Contrast this with his previous allusions to "marking" (*orizei*) *kinêsis*. Before moving on to discuss the relation of soul to time, then, I want first to say something about Aristotle's use both of *arithmos*, or number, and *metron*, or measure, in his various definitions and explanations of time leading up to this discussion.

In the Treatise on Time, Aristotle uses three different verbs to describe the apprehension of time and their corresponding nominal forms to refer to that which time is. He says that *kinêsis* is counted, *arithmêton*, measured, *metrêton*, and marked, *orizei* (see 219a22 "we have marked motion," 219a25 "we mark them," and 220b15, "time marks the movement"). But, as just mentioned, *orizei* is not synonymous with either *arithmêton* or *metrêton*. Because Aristotle uses both *arithmos* and *metron* in the time section, it has been argued that he uses them interchangeably (see Annas 1975, 99). Since *metron*, literally "that by which anything is measured," seems to be a genus of various kinds of "thats," it has also been argued that Aristotle understands number in this case to be a kind of measure (see Coope 2005, 100). In *Metaphysics* x 6, Aristotle explains that, "Plurality is as it were the class to which number belongs; for number is plurality (*plêthos*) measurable (*metrêton*) by one" (1057a3). This passage has been used not only to defend each of the opposing views above, but also to say that for Aristotle, it is one, as opposed to number, which is under the genus of "measure" (see Klein 1969, 108).

---

[7]Ἄξιον δ᾽ ἐπισκέψεως καὶ πῶς ποτε ἔχει ὁ χρόνος πρὸς τὴν ψυχήν, καὶ διὰ τί ἐν παντὶ δοκεῖ εἶναι ὁ χρόνος, καὶ ἐν γῇ καὶ ἐν θαλάττῃ καὶ ἐν οὐρανῷ. Ἢ ὅτι κινήσεώς τι πάθος ἢ ἕξις, ἀριθμός γε ὤν, ταῦτα δὲ κινητὰ πάντα (ἐν τόπῳ γὰρ πάντα), ὁ δὲ χρόνος καὶ ἡ κίνησις ἅμα κατά τε δύναμιν καὶ κατ᾽ ἐνέργειαν; πότερον δὲ μὴ οὔσης ψυχῆς εἴη ἂν ὁ χρόνος ἢ οὔ, ἀπορήσειεν ἄν τις. Ἀδυνάτου γὰρ ὄντος εἶναι τοῦ ἀριθμήσοντος ἀδύνατον καὶ ἀριθμητόν τι εἶναι, ὥστε δῆλον ὅτι οὐδ᾽ ἀριθμός. Ἀριθμὸς γὰρ ἢ τὸ ἠριθμημένον ἢ τὸ ἀριθμητόν. Εἰ δὲ μηδὲν ἄλλο πέφυκεν ἀριθμεῖν ἢ ψυχὴ καὶ ψυχῆς νοῦς, ἀδύνατον εἶναι χρόνον ψυχῆς μὴ οὔσης, ἀλλ᾽ ἢ τοῦτο ὅ ποτε ὂν ἔστιν ὁ χρόνος, οἷον εἰ ἐνδέχεται κίνησιν εἶναι ἄνευ ψυχῆς. Τὸ δὲ πρότερον καὶ ὕστερον ἐν κινήσει ἐστίν· χρόνος δὲ ταῦτ᾽ ἐστὶν ᾗ ἀριθμητά ἐστιν.

The potential for equivocation on "measure" runs parallel to the potential for the equivocation on "number"; for, as Aristotle himself points out about "*arithmos*," measure can mean both the unit of measure, i.e., the "that," or the measurement itself (see 219b where Aristotle says that number can mean both the number counted and the number with which we count). In the first case, we are talking about "one," and in the second place we are talking about a plurality measured by one. For Aristotle, time is number in so far as it is that which is counted—the plurality and not the one. The impulse to think that the analogous sense of *arithmos* and *metron* are not synonymous here has to do with the idea that Aristotle understands time to be an ordering and not a quantity (see Coope 2005, 104). While I would not have a problem acceding to the claim that there is a non-temporal ordering going on between anteriority and posteriority, it seems important to understand these positions as designating a relation. Yes, relations can connote an ordering, but the fact that such a relation exists does not automatically prohibit that the terms in relation, the relata, exist as a discrete plurality or quantity of things. I thus maintain the standard view that number and measure are synonymous in Aristotle's treatment, on the basis that order and quantity are not mutually exclusive designations, and I understand them both to refer to the plurality counted and not the unit, one, by which we count.

With that said, we return to the passage on time and the soul. Recalling the first few lines from the passage above, Aristotle introduces the topic with a statement and a quasi-question, he thinks it "is worth considering how time can be related to the soul (ψυχή); and why time is thought to be in everything (ἐν παντὶ), both in earth and in sea and in heaven." Aristotle wants to consider how time is related to the soul, here not yet qualified as the rational soul. Time is thought to be in everything, meaning in things on earth, in the seas, and in the heavens. Though, since Aristotle has offered an unconventional definition of time here in the Treatise on Time, the idea commonly held that time is "in everything" is right, but now in a new sense. For Aristotle, time is in everything because, (1) "it is an attribute, or state (πάθος ἢ ἕξις), of movement (κινήσεώς) (since it is the number of movement)," and (2) "all these things [on earth, in the sea, and in the heavens] are movable (for they are all in place), and time and movement are together, both in respect of potentiality and in respect of actuality." If time is the number of *kinêsis*, it is not an intrinsic part of natural objects. Indeed, as I have argued, it has no existence for Aristotle qua itself and unless actualized remains a potentiality of *kinêsis*.[8] Yet, to the extent that natural beings on earth, in the sea, and in the heavens, undergo *kinêsis*, and *kinêsis* is an actualized potentiality because they are first of all actually in place, there is the potentiality for these natural beings to be in time.

Since at this point Aristotle has said only it is worth considering that time is related to the soul and clarified that time is an attribute of *kinêsis* because it is a

---

[8]See Polansky (Polansky 2007, 463 n. 5) on interpretation of *hexis*. For Polansky, the examples of light and art in *De anima* iii as *hexis* provide support that *hexis* can mean potentiality or disposition. It seems that *chrónos* as *hexis* provides further evidence that *hexis* is a potentiality for possible actualization under certain conditions.

number of *kinêsis*, the specific relationship of time to soul is not clear, but it does seem clear that it is going to have something to do with the sense in which time is a number, and number, as we saw previously, is something counted. A question thus could be raised as to whether this counting is done not by anyone in particular, but in accordance with some celestial standard, as it has been argued, or if it results from direct observations and then counting of *kinêsis*. This difference is parallel to the question raised in the previous chapter regarding whether Aristotle's analytic of time was an analytic of infinite time or time taken. It is worthwhile to address the analog to the previous question we find here. Understanding time as the number in accord with a celestial standard annihilates the possibility that time is actualized by the interaction between the observed and the observer and so too my previous claim that Aristotle is focused here in the *Physics* on the time taken. Instead, time becomes something a priori, namely, what we might take to be infinite time, unaffected by particular instances of existing self-subsistent natural beings undergoing *kinêsis*.

In addition, it seems suspicious that Aristotle would argue for the definition of time that he does, if he just meant to explain time as a pre-established standard—essentially predetermined before any *kinêsis* takes place and unalterable by particular *kinêsis* and observation. Certainly, given the context of his scope, access, method and goals in the *Physics*, it is unclear as to why, if time were really just a set number naming the perfect motion of the heavens, it appears in this context at all.

Returning again to the text, Aristotle asks another question, which at this point seems redundant, namely: "whether if soul did not exist time would exist or not." But, now we get an explicit answer, "if there cannot be some one to count there cannot be anything that can be counted either." Whereas someone counting is not requisite so that "anything" exist, it is requisite in order that "anything" be counted. Aristotle here makes a general claim about the relationship between things existing, things being counted, and someone counting. Whereas, the claim that something counted, i.e., number, depends on someone counting may seem like a strange claim (one generally accepts that there can be eight planets in the solar system whether or not they are ever counted), the ideas of counting (by way of the counter) and the counted are intimately related in ancient Greek.

It has been argued that our modern concept of number, which comes from Descartes and Leibniz, is vastly different from the concept of number employed here by Aristotle (see Sachs 2008, 129). In Greek mathematics, numbers are names given to a discrete plurality of things (see *De interpretatione* ii on names as convention). They are "natural" and not symbolic expressions (see Sachs 2008, 130 and Klein 1969 regarding fractions and negative expressions). Again, Ross (Ross 1936, 541) explains in reference to *Meta* 987b27 that "the Pythagoreans identified real things with numbers, it is not to be supposed that they reduced reality to an abstraction, but rather that they did not recognize the abstract nature of numbers" (see also fn 47 in Chap. 1). While the plurality of things to be counted exists outside of the fact of someone's counting them, the name given to the plurality is only potentially so. In order for number, as name, to arise, the plurality—the something to be counted—must be apprehended, thus named. In the case of time, as we know, the something to be counted is *kinêsis* to the extent that this is the mode of existence for natural beings. The sense in which *kinêsis* becomes numbered, and thus the sense in which time exists at all on this account, has to do with whether or not there is someone counting it. Indeed, since on Aristotle's definitions, time is a number, and "number is either what has been, or what can be, counted," number is arrived at by way of counting. It is thus implied that someone or something is doing the counting. Aristotle's claim here is that the number, i.e., time, necessarily depends on the counter.

It is the "some" of this someone counting—namely, who or what is it—that has caused so much debate over this passage in Aristotle's Treatise on Time. From the first section above, this someone could not be any ensouled being, i.e., plants, nonhuman animals, and humans alike. Unlike the act of simply marking (*orizei*), counting—really a type of naming—seems to be uniquely human. Looking back to the passage, Aristotle seems to say as much: "But if nothing else is of such a nature as to count but the soul and the intelligence (*nous*) in the soul. Then it is impossible that time be if soul is not, but only that of which time is an attribute." The actual existence of time, then, requires not simply soul, as it is often suggested and consequently misunderstood, but the intellective capacity of soul, or *nous*. It is the intellective faculty of the human soul that allows for a readiness for counting or naming, a potentiality, that is not present either in Aristotle's definition of God or in the souls of nonhuman beings (compare with *De anima* iii 4 "And indeed, they speak well who say that the soul is a place of forms, except that it is not the whole soul but the intellective soul, and this is not the forms as being-fully-itself, but in potential" 429a). Time is actualized when a human being with readiness for thinking brings this potentiality to bear on a being actually undergoing *kinêsis*.

Aristotle concludes the passage with a reminder about what is actually being counted: "The before and after are attributes of movement, and time is these *qua* countable." Whereas, I have emphasized before the notion of "marking" the difference from "before" to "after," thus not quite counting, here Aristotle uses the term *arithmêton* instead of *orizei*. One wonders how and/or why the "before" and "after" are sometimes marked, and marked by some nonhuman animals, and yet sometimes counted, seemingly only by human beings, i.e., those with *nous*.

Again, returning to time's identity as *number*, it is something about the number, which allows for the disparity in Aristotle's language about time apprehension. But, what is it about numbers, which could allow for the lack of congruence we see in Aristotle's descriptions of their apprehension? Numbers, as referents for discrete quantities of real things, instead of self-subsistent beings themselves, do not have attributes (recall 204a8–29, number is not a substance). Thus, it must be the number itself, i.e., the quantity of things *it* names, which makes the difference for its potential cognition. When the being in question undergoes only smallscale *kinêsis*, here understood to mean a difference between the "nows,"—this one, "before," and this one, "after"—not separated either by an extended spatial continuum or by many intermediate "nows," the time or number of the *kinêsis* can be perceived. For example, if I walk across the room, the before and after of the locomotion is apprehended easily by another animal in the room. Here, perception of change seems to allow for a rudimentary or partial perception of time. The number of the change is so small that it does not seem to require counting. When the locomotion happens over a greater spatial magnitude, and thus apprehending it requires recognizing what becomes a continuum of change over the magnitude, e.g., I start in New Haven and end up in Thessaloniki, a more robust faculty for apprehension appears requisite. The change is too great to mark, and indeed I wager that no nonhuman animal (or small human child)[9] measures precisely such a change—they certainly detect a difference between places (something changed!), but not the change itself, thus not the numeral of the change.[10] This explains why Aristotle reintroduces the term *arithmêton* when he discusses the relationship between time and the soul and then clarifies *nous* as the additional faculty necessary to apprehend the time.

It seems appropriate then to distinguish between time perception based solely on sensation, which seems to be the course of perceiving and "marking" (*orizei*), as we saw in the previous chapter with our discussion of *Physics* iv 11, and time perception made more precise by the capacity for enumeration (*arithmêton*). This is to say that the potential for time exists in all *kinêsis*, and it is in some sense recognized by the sensitive soul, but the rational faculty of the soul is required in order to bring time, at least in the case where the number that time names is a large quantity of discreet beings, from a hazy multiplicity to a known quantity. Counting sets humans apart from nonhuman animals. We can differentiate a multitude by counting. This allows us to move past sensing number, hence employing our souls' intellective potency to determine the discrete number of items that we sense to be a multitude. Thus, counting looks to require both a body as medium for obtaining sense data and a higher order intellect to discern number. Counting motion, which amounts to the coming into actuality of time, then, requires living beings capable of sensing the before and after in motion and, when we are not just dealing with short-term *kinêsis* or a small quantity of discrete existing beings, a readiness for intellection in order to number, or name, the plurality. Aristotle, then, leaves taking time, generally, to human being. But, he allows that nonhuman animals perceive small-scale change and time, without which they would not have the tools to serve necessary ends, e.g., the capture of prey and evasion of predators. In the next and final section, I will offer up evidence, from Aristotle's *Parva Naturalia* and biological treatises, supported by experimental results from contemporary science, in further defense of this position.

---

[9] See for example *HA* 588a24–588b6, where Aristotle equates the psychology of a child to that of a nonhuman animal.

[10] King (2009, 62), also distinguishes between perceiving and measuring time. His argument is that, "representations are necessary to the perception of common perceptibles such as change and magnitude, and also for the cognition of time. Because representations are a function of perception, this means that time is perceived." He notes that Aristotle (echoing Irwin) does not mention memory in the *Physics*. He concludes, "representation is responsible for the perception of time." defended by 450a9–12, where Aristotle says, according to King, "it is necessary for change and magnitude to be perceived with the same thing that time is perceived."

3.3 Perceiving Time Revisitied

In *De* sensu *et sensibilibus*, Aristotle takes up discussions that would have been too specific for his general work on the soul, *De anima*. He refers to these as the "remaining part of our subject" where he means specifics about soul. Here, we are going to get into the details of soul functioning. Despite that we learn in *De anima* about certain faculties of soul, which do not require the body as medium, the soul never functions disembodied. In *De* sensu *et sensibilibus*, Aristotle's topic turns to a more focused discussion of what he names the most common and important faculties of soul, those that require both soul and body. He explains that these faculties—sensation, memory, passion, appetite, desire, pleasure, and pain—belong to all animals (436a8–10). Indeed, they can be tested to reveal that both soul and body are necessary for their proper operation. One does not see without an eye, but neither does a corpse or a brain-dead animal even with eyes. The brain in the vat does not feel pain, but neither do the disemboweled organs. Aristotle reasons that this is the case because these faculties "all either imply sensation as a concomitant, or have it as their medium" (πάντα γὰρ τὰ μὲν μετ᾽ αἰσθήσεως συμβαίνει, τὰ δὲ δι᾽ αἰσθήσεως); he then concludes that sensation is a faculty of soul inextricable from the body through which external stimuli are taken in (436b1–9).[11] He continues, explaining that while the senses are a natural attribute of the beings (Cf. *HA* 533a15–18), which we call "animal" (*zoon*); indeed, it is by the faculty of sensation that "we distinguish between what is and what is not animal" (ἀνάγκη ὑπάρχειν αἴσθησιν· τούτῳ γὰρ τὸ ζῷον εἶναι καὶ μὴ ζῷον διορίζομεν) (436b11–13); they operate for different functions in different animals.

For Aristotle, despite that sense perception is activity (*energeia*), which is an end in itself, the senses are also a means to an end, and the ends (*teloi*) differ in accordance with the varied potencies of souls for which he argued in *De anima*. This difference is seen first with regard to the senses requiring an external medium to operate: smelling, hearing, and seeing (436b18–19). We are told animals that move locally possess these senses, and for all of them these senses are means for basic survival. Animals can use smell, sound, and sight to find food and/or to be alerted to possible dangers. But, these senses can, "…serve for the attainment of a higher perfection. They bring in tidings of many distinctive qualities of things, from which knowledge of things both speculative and practical is generated in the soul" (τοῖς δὲ καὶ φρονήσεως τυγχάνουσι τοῦ εὖ ἕνεκα· πολλὰς γὰρ εἰσαγγέλλουσι διαφοράς, ἐξ ὧν ἥ τε τῶν νοητῶν ἐγγίνεται φρόνησις καὶ ἡ τῶν πρακτῶν) (437a1–4). These higher ends are restricted to animals that have intellect (τῶν νοητῶν), i.e., to humans.[12]

---

[11]ὅτι δὲ πάντα τὰ λεχθέντα κοινὰ τῆς τε ψυχῆς ἐστι καὶ τοῦ σώματος, οὐκ ἄδηλον. πάντα γὰρ τὰ μὲν μετ᾽ αἰσθήσεως συμβαίνει, τὰ δὲ δι᾽ αἰσθήσεως, ἔνια δὲ τὰ μὲν πάθη ταύτης ὄντα τυγχάνει, τὰ δ᾽ ἕξεις, τὰ δὲ φυλακαὶ καὶ σωτηρίαι, τὰ δὲ φθοραὶ καὶ στερήσεις· ἡ δ᾽ αἴσθησις ὅτι διὰ σώματος γίγνεται τῇ ψυχῇ, δῆλον καὶ διὰ τοῦ λόγου καὶ τοῦ λόγου χωρίς.

[12]Cf. *GA* 731a30–731b7: sense perception is a kind of knowledge and *HA* 588a24–588b6 where Aristotle claims that there is an analogue for knowledge, wisdom, and sagacity in nonhuman animals and then admits that it is difficult to demarcate human animal from nonhuman animal potentiality.

Yet, whereas the distance senses of seeing, hearing, and smelling allow animals to sense proper sensibles, i.e., that which can be sensed only by being seen, that which can be sensed only by being heard, and that which can be sensed only by being smelled, we learn also of common sensibles (see *De anima* ii 6 for a parallel account). When things can be perceived with more than one faculty of sense, they are sensed in common. Aristotle provides the following list: figure, magnitude, motion, rest, and number. Sight allows us the most variability in sensing, and it plays an especially big role in perceiving common sensibles.

Now, these passages leave us with a lot to think about regarding the way sense perception functions to allow animals—both human and nonhuman—to attain various ends. Both humans and nonhumans, in so far as they are capable of locomotion, can see, hear, and or smell. But, what can they see, hear, or smell? In Aristotle's biological works, we find myriad examples of nonhuman animal perception. Consider, for instance, these passages from the *Historia Animalium*: Fishes are repelled by loud noises (533b4–534a7), e.g., those that seek shelter in holes after hearing men rowing; dolphins beaching themselves as a response to loud splashing[13]; shoals of fish scurrying away at the slightest sound; sub-rock dwelling fish that emerge when stones are clashed against the rock. In these examples, Aristotle attributes animal action to the animal's sense of hearing. This explains his amazement that the fish hear without any clear instrument for apprehending sounds, and likewise that they seem to smell without an instrument for olfactory perception (533b1). In each of the cases given, however, the animals seem to be detecting motion, a common sensible. As a common sensible, they could be hearing the motion or sensing it by some other means, e.g., by touch—feeling the vibrations of the clanging rocks or the splashes in the water made by the oar. In any case, it is clear that in these examples, perception of motion functions to effect consequent movement and action.[14] The resulting movement serves greater ends, such as attempt at preservation of life.

Aristotle provides other concrete examples of animal sense perception; he tells us that the octopus will relinquish its unusually firm grip on rocks at the first smell of fleabane (534b26–30), that the hyaena will await a passerby in order to prey upon him and—from another perspective, that the dog will fall prey to the hyaena when persuaded by its strange vocalizations mimicking a vomiting man (594a32–594b4). He tells us also of the enmity between the horse and the anthus. Aristotle states unequivocally that the bird sees poorly. It thus relies on its sense of hearing to, as Aristotle explains, mimic the horse's vocalization and fly at the horse to persuade it to leave—its only defense against the horse's nefarious intentions (609b15–19). Hereto, we see perception effecting movement and action in the service of other ends.

---

[13]Aristotle infamously classified dolphins as fish, specifically "dualizers."

[14]On the causes of movements and actions in animals, see *De motu* vii, especially 701a25–35 and viii, especially 701b34–702a6. According to Aristotle, regarding animal movement: "… the proximate reason for movement is desire, and this comes to be through sense-perception or through *phantasia* and thought (Τῆς μὲν ἐσχάτης αἰτίας τοῦ κινεῖσθαι ὀρέξεως οὔσης, ταύτης δὲ γινομένης ἢ δι'αἰσθήσεως ἢ διὰ φαντασίας καὶ νοήσεως)" (34–35).

In the second example, the end is attainment of nourishment, and in the third example, the end is self-preservation—most critical aims! That Aristotle claims the sheep to be the least intelligent of the quadrupeds because it leaves its herd and/or shelter for no reason—often to its own demise—is further confirmation that sensation for Aristotle is not only active in nonhuman animals, but that it is meant to be used in service to a purpose (610b21–25). When the animal moves locally without an aim, especially when such action goes against self-preservation, it is said to lack intelligence. We learn as well of the owl and the night-raven, who, opposite fish (602b5–9) see poorly in daylight (619b19) and at length of the highly intentional life of bees, who, as it were, are put off by malodours (623b5–627b23). We see in these examples that insofar as perception is often useful for animals, i.e., in service to important ends, we might then consider how perception of motion and number serve such higher ends. Could animal perception without intellection allow for some sort of time apprehension?

Number is also a common sensible, according to Aristotle. Insofar as number is typically perceived by way of enumeration, and not sense perception, one wonders what Aristotle is up to here. How are numbers perceived by the sensitive soul, and to what ends? In the context of our conversation of time, where time is classified as a number—but, a number *derived of* motion—it is likewise pertinent to ask about the possibility that time can be actualized, i.e., the number of motion can be apprehended, only by way of perception. Yet, while the answers to these queries are not directly answered in the Corpus Aristotelicum, the idea that at least some time can be perceived without enumeration is consistent not only with the examples given above, where nonhuman animals are perceiving motion, but also with Aristotle's language beginning in *Physics* iv 11, where he talks about time apprehension as perception and marking and with his arguments in the treatise on memory. Our final task, then, is to inquire as to how perceiving number and counting number differ in the service of time apprehension. In an effort to present Aristotle's claims, both that time requires *nous* and also that (1) nonhuman animals have a sense of time, (2) both motion and number are common sensibles, and (3) time can be perceived (and motion marked), as consistent, I will incorporate conclusions from contemporary science to support the claim that while in fact some numbers, i.e., small numbers typically less than four, can be perceived even by infants and nonhuman animals, larger numbers must be enumerated. Such evidence lends hard proof to Aristotle's insights about the complexity of "taking time."

Given our common experiences with perception, we understand that even humans seem to sense only small numbers. When I see two apples on the table, for example, I can say without thinking that there are two there. When there is a bushel on the table, however, I can only immediately say that there are many. I would have to count them to know exactly how many are there. When I hear three notes strum on a guitar, I seem to hear them without enumerating them; but, when many notes are strum in quick succession, I can no longer discern how many there have been. Indeed, experimental programs in psychology and neuroscience know this to be the case. According to Kaufman et al.'s landmark study (1949), whose conclusion effectually synthesizes the two prevailing yet seemingly opposed conclusions at their time; there is no immediate and adequate perception of number,[15] but there is an activity whereby numbers six and under are rapidly and accurately discriminated. They name this activity, "*Subitizing*," from the Latin "subitus," sudden. Trick and Pylyshyn (1994) confirm that small and large numbers are enumerated differently; they accept Kaufman et al.'s term and, further, conclude that subitizing relies on preattentive information, whereas counting requires spatial attention. In a recent study, Harvey et al. (2013) conclude that, "numerosity perception resembles primary sensory perception and, indeed, it has been called the number sense"; and, "the cortical surface area devoted to specific numerosities decreases with increasing numerosity."[16] But, even if human adults subitize quantities of about six and under, what evidence do we have that nonhuman animals and human children do the same?[17] Do nonhuman animals and human children have the information Trick and Pylyshyn consider preattentive?

Let us return first to the examples of nonhuman animal perception of motion from Aristotle's biological works. It seems clear that nonhuman animals sense number, even if the exact quantity remains unknown. Thinking again about the hyaena, one would not say that if the hyaena should encounter twenty men or fifteen dogs, instead of one in each case, that she is somehow unaware that there are multiple. In order to catch one, she not only sees the many, but also devises a strategic plan for isolating her anticipated catch. Aristotle observes that when the lion is pursued by many men at once, his behavior is different than when he is either not being pursued or when he is himself pursuing other prey (629b14–20). These examples tell us that nonhuman animals, on Aristotle's account, do alter behavior when faced with multiple, as opposed to one, objects. Returning to my previous point that "*nous*" in *Physics* iv 14 must mean the working *together* of sense and intellection in that the faculties of intellect require sensation, here we see the possibility for limited time apprehension insofar as the number is small enough to sense, even if the potency of intellect is not present to calculate or measure.

In fact, recent research in neuroscience, psychology, and animal cognition confirms Aristotle's observations, showing that many animals at various ages perceive number without counting (see for example Cooper et al. (2003, 236), dogs seemed to have "some concept of number of objects, though it would not be fair to infer anymore than simple subitising of number. It may therefore be that dogs only represent numbers of objects as 'one,' 'two,' and 'lots.'" Further, Dormal et al. (2006, 110) report:

---

[15]Cf. again Klein 1969 on the possible intuitive nature of *arithmos*.

[16]Reas (2014) reports these results for laymen in her recent review of this study: "One side of this brain region responds to small numbers, the adjacent region to larger numbers, and so on, with numeric representations increasing to the far end.".

[17]Kaufman et al. address this question, admitting that the conclusions in their study are based on a study of adult human number perception. To include children and nonhuman animals in such a study, they suggest that one consider that: Subitizing, Estimating, and counting are all learned behaviors (1949, 524).

Animal data show that various species can discriminate numerosities in experimental as
well as in natural conditions...There is also clear evidence that newborn babies and young
children experience time and have a precocious temporal representation...These elementary
numerical skills shared by animals, infants and adults would rely on a cerebral network
located in the inferior parietal cortex.

Though, whereas these studies were conducted almost exclusively testing
number perception with the sense of sight, Riggs et al. (2006) proved that subsi-
tizing can happen even with what Aristotle calls the most distributed sense, touch
(see Riggs et al. 2006). This opens up basic time perception to all animals, of all
ages, on Aristotle's account.[18] Aristotle's vast experience with the natural world
seemingly led him to these same general ideas about the possibilities for soul
functioning in all animals.

But, of course, time for Aristotle is a number of *kinêsis*, not a static quantity.
How do we see number when the numbers perceived are not all present at once?
Specifically, how can perception of not only number, but the number of motion,
allow for time apprehension in Aristotle? King (2009, 65) rightly points out that
Aristotle does not discuss this point; King's explanation, which I think is correct, is
the result of a contextual approach: Aristotle's "theory of change does not allow for
change or rest at an instant, and also because his theory of time requires the
cognition of change, rather than being itself a presupposition for the cognition of
change." King answers this problem with a theory of representation, based on
perception; namely, he suggests that we perceive representations (images) of the
perceptibles and that the "now" perceived before is held in representation even as
we experience it change to "now" perceived as after. He concludes that, "remaining
representations make it possible to perceive time, which is one of the preconditions
of memory" (66). Indeed, Wood et al. (2008) confirm that rhesus monkeys can
differentiate between small numbers of non-solid portions of food, not poured
simultaneously, up to the number four and Agrillo et al. (2008): mosquitofish are
able to discriminate between small numbers of non static objects, notably as high as
the difference between three and four. West and Young (2002) show that nonhuman
animals can understand simple calculations, e.g., two treats are shown to a dog and
one treat is taken away; the dog notices the difference between two and one.

Here, we see contemporary science providing experimental evidence for the
conclusions Aristotle seemingly developed about time perception—especially, in so
far as it is a type of number that, if perceived, must be perceived with motion or
change, the possibility that it happens with nonhuman animals. Aristotle's observations about animal behavior, as presented in his biological works, demonstrates
that there would have been a conflict between his conclusion at *Physics* iv 14, that
readiness for thinking would be required for time apprehension, and his explicit

---

[18]I am grateful to Lanei Rodemeyer and Heidi Lockwood, for sharing with me on separate
occasions that their infant children seem to experience time. And, indeed, we see here that
scientific results confirm that they do. The question remains as to the extent and nature of their time
apprehension; I propose that Aristotle was correct to group human children with animals in so
far as they are likely using faculty of sense as opposed to thought to cognize temporality.

claim in the *De sensu et sensibilibus* about nonhuman animal time apprehension. This tension is confirmed by his varied language in the Treatise on Time, switching among language indicating that time is apprehended by way of the perceiving, marking, counting, and measuring of motion. Certainly, extended change, e.g., the trip from Thessaloniki to Athens, or the growth from infancy to adulthood, is not going to be cognized by nonhuman animals. There is simply too much change to keep track of—in King's language, to represent—and for which to account, and, given Aristotle's understanding that time is a number, thus too great a number to perceive. When the number is a small quantity, corresponding to a small scale change, e.g., the lion runs across the field, time in Aristotle's understanding would be easily cognized without a readiness for thinking—the lion was at one end of the field, and now he is at the opposite end—so the number of the change is small enough to perceive without any further activity of soul. This conclusion is widely supported by Aristotle's examples of nonhuman animal behavior in his biological treatises, e.g., to give a negative account, if such changes were *not* cognized, intentional necessary ends, such as catching of prey and avoiding predation, would be impossible.

In so far as nonhuman animals perceive, and Aristotle allows for perception of time, it seems likely that, for Aristotle, sense perception is a necessary but not sufficient condition for full time sense. Time is a number for Aristotle; to the extent that many, if not all, animals can mark (*orizei*) before and after in *kinêsis*, sense-perceiving number, they must have a weak sense of time, i.e., perception of small-scale change. But, insofar as the number must be counted or measured, time actualization seems left to humans (see Ross 1936, 599 on *orizei*, that it is not the same thing as measuring time). It is the intellective faculty of the human soul that allows for a readiness for counting or naming, a potentiality, that is not present either in Aristotle's definition of God or in the souls of nonhuman beings (compare with *De anima* iii 4 "And indeed, they speak well who say that the soul is a place of forms, except that it is not the whole soul but the intellective soul, and this is not the forms as being-fully-itself, but in potential" 429a). Time in any case is actualized when a human being with readiness for thinking brings this potentiality to bear on a being actually undergoing *kinêsis*. Human beings, as beings with both sensitive and rational souls, are thus a sufficient but not necessary condition for partial time sense and, along with the natural being undergoing change, both the necessary and sufficient condition for full time actualization in Aristotle's account.

Concluding Remarks

In pursuit of the nature of time (*chrónos*) in Aristotle's natural philosophy, I started out talking about the wider context of his Treatise on Time (*Physics* iv 10–14), taking first his foregoing arguments in *Physics* i–iv 9 and coming subsequently to relevant arguments from other works in his philosophy of nature. Aristotle's goal to understand the nature of natural beings brought him from discussing the natural beings themselves to topics derived from the way natural beings exist in the world, i.e., their nature is an inner potentiality for *kinêsis*. Since Aristotle's study of time comes from his interest in nature and time for him is not an existing self-subsistent natural being to investigate *qua* itself, but something "taken," it has been difficult for readers of Aristotle to know exactly how to understand what time is for Aristotle. This is particularly the case when we look at other works in Aristotle's natural philosophy, which add additional details about what he understood time to be.

There is something ephemeral about time in that, as Aristotle puzzles about in *Physics* iv 10, it does not really seem to exist. This peculiarity, as we saw, is characteristic of all terms of *kinêsis* for Aristotle. The sense in which time seems clearly to exist, and yet can be said really not to exist has to do with Aristotle's interest in the modality of potentiality in nature—an interest, I have argued, which defines his natural philosophy and sets it apart from his predecessors. As we saw in the cases of the infinite and place, time is only ever potentially existent—only ever potentially a continuum and a whole with parts—in so far as it is derived from that which does actually exist in this way. While place can become actual when there is a natural being occupying magnitude, the infinite and time both require something beyond the materiality of natural beings in order that they be actualized. Infinity is the potentiality for endless divisions of beings that never actually occurs. The sense in which the infinite exists is as an actualized thought about the possibility for continuous beings. It is a consequence of continuity recognized by the rational soul. Similarly, time only ever becomes continuous itself as an actualized attribute of *kinêsis*, when a change in "now," referring to the natural change of "thisis" or substantial natural being, is apprehended and marked/counted/measured.

This is not to suggest, however, that humans are divorced or excised from nature. Humans are certainly natural objects themselves on Aristotle's account, but they are unique natural objects in that they have a divine-like faculty, i.e., the

© The Author(s) 2015
C.C. Harry, *Chronos in Aristotle's Physics*,
SpringerBriefs in Philosophy, DOI 10.1007/978-3-319-17834-9

rational soul. Time, then, is not something to discover, or learn about, which explains why it receives relatively little attention in Aristotle's corpus. Instead, Aristotle recognizes it as something we use to make sense of things; by way of actualizing time in the derivation of our apprehensions of nature, we better understand our relationship to other natural beings—perhaps, in so far as we are able to perceive our own inner motions, so too our relationship with ourselves.

In an effort to understand Aristotle's thought as internally consistent as regards his claims both that time requires readiness for thinking, on the one hand, but also that time can be perceived, on the other hand, the idea that nonhuman animals perceive time was explored. Indeed, examples from Aristotle's biological works together with conclusions from contemporary scientific research, suggest that a weak version of time perception, i.e., the apprehension of small multiples, is possible for humans and nonhumans using only the sensitive soul.

Does my reading of Aristotle's Treatise on Time commit him to the view that, if there is no one at least to perceive motion, then there is no actual time? The short answer is, yes. This is not, however, a problem per se. The point is this: substantial natural beings exist independently of perception; likewise, the way these beings exist, as habitually changing, exists independently of perception. But, what does not actually exist independent of perception (and, in many cases, intellection) is time, viz., the number of before and after with respect to change—what I have here called a product of interaction between various beings in nature.

Bibliography

I. Works by Aristotle in English, Unless Otherwise Stated

J. Barnes, ed. 1984. *Revised Oxford Translation (ROT) of the complete works of Aristotle*, Vol. 2. Princeton: Princeton University Press.

II. Selected Unpublished Works from the 5th International Symposium: The Issue of Time in Aristotle, Thessaloniki and Naoussa-Mieza, 2012 (Forthcoming in Publication with VRIN)

Harry, Chelsea, and Ron Polansky, "When time is accidental".
Heinemann, G. "Time as measure".
Keizer, Helena. "AIΩN and time in Aristotle".
Puente, Ray. "Selon Aristote, perçoit-on ou pense-t-on au maintenant, au temps et au nombre?".
Sfendoni-Mentzou, Demetra. "Is time real for Aristotle?".

III. Other Works

Agrillo, et al. 2008. Do fish count? Spontaneous discrimination of quantity in female Mosquitofish. *Animal Cognition* 11(3): 495–503.
Annas, Julia. 1975. Aristotle, number, and time. *Philosophical Quarterly* 25: 97–113.
Aquinas, St. Thomas. 1963. *Aquinas's commentary on Aristotle's physics*. Trans. R. Blackwell, R. Spath, and W.E. Thirkel. New Haven: Yale University Press.
Bolotin, David. 1993. Continuity and infinite divisibility in Aristotle's physics. *Ancient Philosophy* 13: 323–340.
Bolotin, David. 1997. Aristotle's discussion of time: an overview. *Ancient Philosophy* 17: 47–62.
Bostock, David. 1980. Aristotle's account of time. *Phronesis* 25: 148–169.
Broadie, S. 1982. *Nature, change, and agency in Aristotle's physics*. Oxford: Clarendon Press.
Bury, R.G. Trans. 2005. Timaeus. In *Plato IX of the Loeb Classical Library*, ed. Jeffrey Henderson. Cambridge: Harvard University Press.
Chase, Michael. Trans. 2006. *Pierre Hadot's the Veil of Isis: An essay on the history of the idea of nature*. Cambridge: The Belknap Press of Harvard University Press.
Coope, Ursula. 2005. *Time for Aristotle: Physics IV. 10–14*. Oxford: Clarendon Press.
Cooper, J., et al. 2003. Clever hounds: social cognition in the domestic dog (Canis Familiaris). *Applied Animal Behaviour Science* 81: 229–244.

© The Author(s) 2015
C.C. Harry, *Chronos in Aristotle's Physics*,
SpringerBriefs in Philosophy, DOI 10.1007/978-3-319-17834-9

Couloubaritsis, Lambros. 1997. *La Physique D'Aristote*. Bruxelles: Ousia.

Curd, Patricia ed. 1996. *A presocratic reader: Selected fragments and testimonia*. Trans. Richard McKirahan. Indianapolis: Hackett.

Danielson, Dennis Richard. 2000. *The book of the Cosmos: Imagining the world from Heraclitus to Hawking*. Cambridge, MA: Perseus.

De Moor, Mieke. 2012. *Aristote et la question du temps: avec la traduction française de l'ouvrage de Gernot Böhme, "Zeit und Zahl" introduction, première et deuxième parties relatives à Platon et Aristote*. Dissertation, Université Aix-Marseille.

Dormal, V., et al. 2006. *Acta Psychologica* 121: 109–124.

Ellis, Brian. 2001. *Philosophy of nature*. London: Acumen.

Gabriel, Marcus. 2009. God's transcendent activity: Ontotheology in Metaphysics 12. *The Review of Metaphysics* 250 (Dec 2009): 385–414.

Galileo, Galilei. 1957. Letters on Sunspots. In *Discoveries and opinions of Galileo*, ed. and trans. by Stilman Drake. New York: Anchor Books.

Gower, Barry. 1973. Speculation in physics: The history and practice of Naturphilosophie. *Studies in the History and Philosophy of Science* 3(4): 301–356.

Gottshalk, H.B. 1965. Anaximander's apeiron. *Phronesis* 10(1): 37–53.

Graham, Daniel. 1988. Aristotle's definition of motion. *Ancient Philosophy* 8: 209–215.

Harvey, B.M., Klein, B.P., Petridou, N., and S.O. Dumoulin. 2013. Topographic representation of numerosity in the human Parietal Cortex. *Science* 6: 341 (6150), 1123–1126.

Heidegger, Martin. 1998. On the essence and concept of φύσις in Aristotle's physics B, 1. In *Pathmarks*, ed. William McNeill, 183–230. Cambridge: Cambridge University Press.

Henry, Devin. 2011. Aristotle's pluralistic realism. *The Monist* 94(2): 197–220.

Hussey, Edward. Trans. and Comm. 1983. *Aristotle's physics, books III and IV*. Oxford: Clarendon Press.

Jowett, Benjamin. Trans. 1961. Timaeus. In *The collected dialogues of plato* ed. Edith Hamilton, and Huntington Cairns. Princeton: Princeton University Press.

Judson, Lindsay (ed.). 1991. *Aristotle's physics: a collection of essays*. Oxford: Clarendon Press.

Kaufman, et al. 1949. *The American Journal of Psychology* 62(4) (Oct): 498–525.

Keizer, Helena Maria. 1999. *Life time entirety: A study of AION in greek literature and philosophy, the Septuagint and Philo*. Dissertation, University of Amsterdam.

King, R.A.H. 2009. *Aristotle and plotinus on memory*. Berlin: De Gruyter.

Klein, Jacob. 1968. *Greek mathematical thought and the origin of algebra*. Trans. by Eva, Brann. New York: Dover Publications.

Koren, Henry J., and C.S.Sp, S.T.D. 1958. *Kleines Lehrbuch des Positivismuus* reprinted in *Readings in the philosophy of nature*. Trans. ed., Intro, and Comm. Richard von Mises. Westminster: The Newman Press.

Kosman, L.A. 1969. Aristotle's definition of Motion. *Phronesis* 14: 40–62.

Mansion, A. 1945. *Introduction à la physique aristotélicienne*. Louvain: Institut Supérieur de Philosophie.

Maudlin, Tim. 2012. *Philosophy of physics: Space and time*. Princeton: Princeton University Press.

Miller, Fred. 1974. Aristotle on the reality of time. *Archiv fuer Geschichte der Philosophie* 56: 132–155.

Moutsopoulos, Évanghélos. 2010. *Reflets et Résonances du Kairos*. Athens: The Academy of Athens.

Nussbaum, Martha Craven. Trans. and Comm. 1978. *Aristotle's De Motu Animalium*. Princeton: Princeton University Press.

Oliveri, et al. 2008. Perceiving numbers alters time perception. *Neuroscience Letters* 438: 308–311.

Owen, G.E.L. 1984. Tithenai ta phainomena. In *Logic, science, and Dialectic*, ed. Martha Nussbaum, 239–251. Ithaca, NY.

Polansky, Ronald. 1983. Energeia in Aristotle's metaphysics IX. *Ancient Philosophy* 3: 160–169.

Polansky, Ronald. Comm. 2007. *Aristotle's De Anima*. Cambridge: Cambridge University Press.

Reas, Emilie. 2014. *Our brains have a map for numbers*. Scientific American.

Riggs, K.J., L. Ferrand, D. Lancelin, L. Fryziel, G. Dumur, and A. Simpson. 2006. Subitizing in tactile perception. *Psychological Science* 17(4): 271–272.

Roark, Tony. 2011. *Aristotle on time: A study of the physics*. Cambridge: Cambridge University Press.

Ross, W.D. 1936. *Aristotle's physics: A revised text with introduction and commentary*. Oxford: Clarendon Press.

Ross, W.D. Trans., Intro., and Comm. 1955. *Aristotle Parva Naturalia*. Oxford: Clarendon Press.

Ross, W.D. (ed.). 1973. *Aristotelis Physica*. Oxford: Clarendon Press.

Sachs, Joe. Trans. and Comm. 1995. *Aristotle's physics: A guided study*. New Brunswick: Rutgers University Press.

Sachs, Joe. 2010. 'Aristotle: Motion and its place in nature' in the internet encyclopedia of philosophy. http://www.iep.utm.edu.

Seidl, Horst. 1995. Beitraege zu Aristoteles' Naturphilosophie. Band 65. Amsterdam: Rodopi Elementa.

Shiffman, Mark. Trans. and intro. 2011. *Aristotle De Anima*. Newburyport: Focus Publishing.

Shoemaker, S.S. 1969. Time without change. *Journal of Philosophy* 66: 363–381.

Simplicius. 1895. *InAristotelis Physicorum Libros Quattuor Posteriores Commentaria*, CAG x, ed. H. Diels. Berlin: G. Reimer.

Smith, John E. 1969. Time, times, and the "Right Time": Chronos and Kairos. *Monist* 53: 1–13.

Sorabji, Richard. 1983. *Time, creation, and continuum: Theories in antiquity and the early middle ages*. Chicago: The University of Chicago Press.

Telford, Kenneth A. Trans. and Comm. 1999. *Aristotle'sphysics*. Binghamton: Institute of Global Cultural Studies.

Trick, and Pylyshyn. 1994. Why are small and large numbers enumerated differently? A limited-capacity preattentive stage in vision. *Psychology Review* (Jan) 101(1): 80–102.

Von Leyden, W. 1964. Time, number, eternity in Plato and Aristotle. *Philosophical Quarterly* 14(54): 35–52.

Waterfield, Robin. Trans. 1996. *Aristotle physics*. Oxford: Oxford University Press.

Waterlow, Sarah. 1984. Aristotle's now. *Philosophical Quarterly* 34: 104–128.

West, R., and R.J. Young. 2002. Do domestic dogs show any evidence of being able to count? *Animal Cognition* 5: 183–186.

Wieland, Wolfgang. 1933. *Die aristotelische Physik: Untersuchungen über die Grundlegung der Naturwissenschaft und die sprachlichen Bedingungen der Prinyipienforschung bei Aristoteles*. Göttingen: Vandenhoeck/Ruprecht.

Index

A
Analytic of time, 32–34, 40, 49, 59
Aristotle, 1–6, 8–31, 33, 35, 37–48, 51–60,
    62–67

C
Context, 19, 22, 33, 44, 59, 64
Continuity, 17, 42, 43
Counting, 51–53, 56, 59–61, 65, 67

M
Method, 1–4, 6–8, 10, 16, 31, 59

N
Natural beings, 1–4, 6, 7, 13–19, 23, 27, 34,
    37, 47, 54, 58, 59
Nature, 2–4, 6, 8, 10, 11, 13–18, 20, 22, 27, 28,
    31, 34, 38, 40, 46, 49, 54, 60
Nonhuman animals, 51–56, 60, 61, 64–67, 70
Nous, 51–53, 60, 61, 65
Now, 6, 9, 11, 14, 20, 28, 36–39, 41–49, 58,
    63, 67

Number, 7, 10, 18, 24, 25, 27, 40, 45, 47–49,
    52, 57–61, 63–67

P
Perception, 3–5, 7, 9, 11, 15, 39, 41, 42, 44, 47,
    52, 54–56, 60, 63–67
Potentiality, 2, 3, 6, 8, 14–16, 18–22, 26, 28,
    43, 52, 54, 60
Psyche, 64, 65

S
Scope, 1, 2, 6, 15, 16, 33, 37, 59
Soul, 30, 36, 42, 43, 51–54, 56–58, 60–62, 64,
    67
Subitizing, 65

T
Taking time, 61, 64
Time, 1, 5, 6, 10, 12, 14, 15, 17, 20, 23, 27,
    32–45, 47–49, 51, 52, 55–62, 64–67

© The Author(s) 2015
C.C. Harry, *Chronos in Aristotle's Physics*,
SpringerBriefs in Philosophy, DOI 10.1007/978-3-319-17834-9

Printed in the United States
By Bookmasters